Beyond Future Shock

ALEX ALANIZ, PH.D.

BEYOND FUTURE SHOCK

A NOVEL

2005

Beyond Future Shock

To My Sparkling Lights, Karen, Maricela And Antonio.

PROLOGUE

You never feel the bullet that kills you." That was the first thing the Baron Karl von Onsager was told by his commanding officer the day he reported to his assigned airfield. "Well that's a relief sir," the young, blond-haired baron had replied with a grin.

A young, aristocratic man of considerable privilege, standing tall, cutting a dashing figure in his aviator's dress uniform and full of bravado, Karl began his military career as a fighter pilot a year past the outbreak of the first world war. From the start, the noble-blooded twenty-three year old, who haled from the highest summits of German society, was liked by all. He had an easygoing demeanor, a wry sense of humor, and was a natural born killer in the skies. Karl also knew how to take care of his friends. With his deep connections, he saw too it that the pilots of his squadron never lacked for the finest champagnes, sausages and cheeses. When things got tough, he would make every effort to help keep his fellow pilots flying as a cohesive band of brothers.

By early 1918, Karl, who was by now an ace many times over, had acquired the rank of lieutenant colonel, but war wearied to the point of dispirited resignation, his famously ebullient disposition had soured. A few weeks afterwards Karl was promoted to squadron commander, the previous squadron commander having died in a fireball while flying over enemy lines. Karl was merely twenty-six years old at the time, but because so many pilots had died by this point in the war, he was considered to be one of his squadron's "old men." Most men would have considered the promotion a great honor. Instead, Karl loathed his new job, most especially his newly acquired duty to write to mothers and widows about the untimely fates of their freshly fallen men.

The day "the" fateful bullet struck, Karl was flying alone amid cotton ball clouds rising high into the timeless atmosphere over a dense, snow-covered forest. Fatigued to the bone after two days of grueling back-to-back missions, and worried about his four year old son Heinreich, Karl was returning to his native Berlin on emergency notice. His wife the Baroness Frieda von Onsager, a tall brunette with blood ties to the Kaiser, had wired him earlier that very morning frantic news that Heinreich had taken gravely ill with high fever, that she feared the worst, and that she wanted him to return home as soon as possible.

The first leg of Karl's return trip was to take three hours, followed by a stop for gasoline, coffee and a sandwich, and a final three hour leg to Berlin. Despite his anxious worries however, the young man's exhaustion began to get the best of him an hour into the first leg. Instead of scanning the skies the way he should have been, the goggled-faced baron began to doze off to the sound of his whirring engine and the gushing slipstream rushing past his blond locks. While Karl fought to keep his eyes open, the clouds drifting past him began shifting into chimeras. One cloud, he noted in passing, was shaped like his father's castle by the Rhine replete with its tall watchtowers, while yet another cloud, hovering just above it, assumed the shape of a fire-breathing dragon.

A bump of turbulence momentarily shook Karl awake. He looked around with half-opened eyes and noticed that he was still alone. Only new clouds surrounded him, and despite his best efforts to stay alert, his eyes closed again. That—the instant his eyes closed—was when all hell broke loose. From out of nowhere, a stream of angry bullets began smashing into his Fokker triwing, and Karl's heavy grogginess at once gave way to surprise and frightening disorientation. A split second later, however, the rain of incoming hellfire had flung Karl back into cold reality—a nasty, years long ordeal of fighting and killing high in the skies that had been too often filled with too many moments of sheer exhilaration or abject terror, or, quiet often, of both simultaneously.

Karl spotted his attacker from the corner of his eye. It was an American flying a French Nieuport 11. The insignia of a Native American head painted on the side of the Nieuport, rather than the French roundels was the giveaway. "Damned American!" Karl cursed. The bastard had apparently been concealing himself high above in the blinding rays of the midmorning sun before suddenly swooping down upon his prey, that being Karl's unsuspecting Fokker triwing flying about a thousand feet below the American. Veering out of the stream of bullets, Karl yanked his plane hard right-about, quickly managing to bank his plane out of harm's way. Only now, feeling like an idiot, did Karl remember the morning's intelligence briefing that an American volunteer Lafayette Flying Corps squadron had been spotted occupying a nearby airfield only the day before. Why, he asked himself, had he so stupidly turned down his wingman's offer to follow him to Berlin. Because he had been thinking about his son and hadn't wanted to wait for Wolfgang to finish his morning duties.

An ace many times over, and ever the aggressor, Karl tried to point his plane's nose upwards to train his own machine guns back at the diving American, but halfway into his maneuver his engine began to sputter, snort and backfire. Literally powerless, Karl was instead forced to break off his counterattack and dive away. From a face full of clouds above him, Karl's eyes and sinking guts now fell upon the snow covered forest rising below his nose. That was when the smell of blood's dank, iron scent caught his attention. Looking down into his cramped, primitive cockpit, Karl was shocked and dismayed. Scattered all about were shredded fibers of muscle meat and splotches of splattered blood. His right leg, apparently, had been riddled with several bullets. Poor Heinreich, Karl thought to himself as he peered down at the gore, I might not make it back home today. Could Karl have foretold of genetic programming, stem cell rejuvenation, brain augmentation with quantum computing bio-nano chips, and other as yet undreamt of technologies that would end death and aging as mankind had known it a mere hundred years hence, and could he have foretold of little, blue-eyed Heinreich's own amazing and critically important future in the so-called solar system mindspace wars that would erupt early in the next millennium, he surely wouldn't have let his haste drop his guard. But Karl knew that now wasn't the time for self reproach. It was time, as his instructors used to say, to assess the situation and take the appropriate action. To that end, with the American now flying below his own plane, trying to recover its altitude, Karl yanked off his grime besmirched goggles and began to peel away at his shredded trousers as gingerly as he could.

Karl nearly vomited at the sight of the gore beneath. The baron's limb was a tattered mess. Bloody, splintered bone fragments protruded from two freshly gouged out bullet holes. The American's bullets had evidently burst into his leg's flesh immediately above his kneecap, and had then ripped through muscle, bone and tendon alike. As Karl looked on with numb astonishment, the two holes filled with dark pools of pulsing, ruddy blood within just a few heartbeats. Two heartbeats later, the warm, blackish red fluid began spilling out at an alarming rate steeping the cushion of his seat a bright rouge.

By instinct Karl knew that if he wanted to live to see his boy Heinreich, he needed to stop the bleeding before doing anything else, and quickly so. The American plane, still turning and climbing hard about below him, wouldn't become a problem for another minute or so. Grabbing his handkerchief, Karl quickly ripped it in two, wadded the individual pieces, and, using his fingertips, began stuffing them into the bigger of the two bullet holes. The larger wound went all the way to his femur. The feeling of touching the bone, splintered as it was, sickened him. Still, resolved to get out the mess he was in, he didn't let the wave of queasiness that gripped him stop him from stuffing more handkerchief into the bleeding hole. When Karl struck a live nerve, he saw a flash of blinding light and then nearly passed out. The pain, which had been mercifully absent so far, now came roaring to life, and with each poke and jab of handkerchief pressed deeper in, the pain seared him like a fire lapping from his toes to his spine and then all the way up to his brain cap. When he finished stuffing the first bullet hole, he gritted his teeth and prayed for the strength to plug the remaining bullet hole. Mercifully, the open tissue of the second wound, due to nerve damage perhaps, remained numb.

With the flow of blood stopped, Karl moved on to deal with his other pending problems, like the American who was now nearly level with him, and the fact that his Fokker was burning at the tail. Because of the growing fire damage, his plane was beginning to buffet to the point of violent jolts and spasms. If the damned thing didn't first shake itself apart altogether, it wouldn't remain flyable for much longer. It was just a matter of time, Karl realized to his dismay, before he would have to bail out.

Looking down at the dense, snow-covered forest which had seemed so peaceful and beautiful moments before, Karl's dismay grew. It was impossible to see through the blanket of white, and in his guts Karl was sure the place was crawling with unseen Brits or Yanks skulking behind every tree, or worse yet, with hordes of ill-humored Frenchmen whom he had bombed and strafed only the day before. That grim thought made Karl wish he could fly on towards safer ground, but with every passing second that he remained in his disintegrating plane, he knew the risk of it trapping him into a death spiral grew increasingly likely. After a futile half a minute of looking for a clearing, or some sign of friendly forces, Karl gave up. Resigning himself to jump, he reached down his flanks, tightened his parachute straps, stood up on his good leg and bailed out of his plane hoping for the best.

Luck, however, was not on Karl's side. At the exact moment Karl jumped, his injured plane rolled itself upside-down, and with an awful, sudden violence the burning carcass batted him hard with a wingtip. Karl was dashed against his plane's fuselage just as the American began shooting at him again. Whump! Whump! Whump! A fresh stream of screaming bullets began slamming into Karl's Fokker and all around his body. A second later the lower starboard wing snapped off in flames, causing the wrecked plane's nose to lurch hard about portside towards Karl's body. The whirring propeller just missed his head before he was bounced away from its spinning blades. The sudden jolt against his injured leg nearly knocked Karl out. Worse yet, as his body started slipping backwards against his Fokker's collapsing fuselage, Karl's shot leg slid into the burning tail section of his plane, whereupon it promptly snagged itself in a tangle of frayed control cables. A fraction of a second later, as Karl continued tumbling backwards past the wrecked tail section, he was suddenly jolted to a harsh stop when the control cables, at the end of their slack and still snagged to Karl's leg, ripped the limb from out of its socket. Luckily, the tendons and muscles held firm and Karl didn't lose his leg. To his utter amazement Karl now found himself dangling backwards by his leg facing the blazing guns of the American plane. With the sound of air and bullets still rushing past him, Karl saw red. Removing his Luger from its holster he screamed out, "I have a wife and child!" at the American and began shooting at him until he emptied his pistol. That one of Karl's bullets actually pinged off the American's propeller greatly impressed the American flyer, who, forced to duck, broke off his attack by banking hard right and climbing away.

Karl, who was not yet ready to give up living, knew he had one last chance. He figured he could use his as yet unopened parachute's tremendous drag to rip him free from his dying plane. Karl closed his eyes, hoped against hope, then pulled his parachute's rip-chord. Two seconds later, with a great snap, rip, pop and another tremendous jolt to his body, Karl, shot leg and all, was indeed torn free. Unfortunately, along with Karl, so too was torn free a large chunk of his Fokker's burning tail. A horrifying second later the burning chunk crashed into his opened parachute, collapsed it upon him, and set the silk ablaze at an edge. Now, without bullets in his pistol to shoot himself with, rather than being pinned in his Fokker to smash into the ground, Karl's fate would be to burn alive long before striking the ground as a simmering corpse. His death would be the long and agonizing death he had most feared. "Please,"

Karl prayed to the American as flames began lapping at his backside, "if you have any mercy, please shoot me."

When the American pilot saw what had happened to Karl, he was horrified, but having shot his first horse after a bad fall as a boy back in Kansas, the man knew what he had to do. Stepping hard on his right rudder, pushing his control stick forward, the American retrained his Nieuport's nose into Karl's plummeting path. It seemed the American, Karl noticed as he tumbled earthward at the edge of panicking, had divined his last wish. At one-hundred yards from Karl's body, the American squeezed his machine guns' triggers. By this point in time, Karl had crossed himself and closed his eyes. Deep within himself, he was somewhere else when the first of the bullets whizzed by his body. He was in Frieda's arms back home in their castle. In his mind's eye, the two of them were standing beside each other, arm-in-arm, happy, smiling, admiring the wondrous beauty of their son Heinreich playing in his room. An instant later Karl heard three more zips, felt a stinging, strangely warm sensation biting into his chest, then a fiery, exploding heat wave erupting inside his lungs, and then he knew of nothing else, for another bullet, smashing into his skull, dispersed his brains to bits in a chunky mist of red and gray spray.

High in the clouds over a peaceful, forested valley, baron Karl von Onsager's bullet riddled body burned to a crisp, parachute and all. A minute after bailing out, the charcoaled lump impaled itself on a sharp, lightning-struck tree stump. It was 10 November 1918. The noble-blooded Fokker ace, with sixty-nine kills to his credit, was gone, killed by a nameless Kansas farm boy flying his last combat mission, for on 11 November, 1918 the Armistice ended the first world war in Europe.

It's funny how insidious change can be Karl had thought after realizing he had been seriously wounded. In one moment he had been a healthy young man. A brief moment later he had become a bullet-riddled wreck, then, not three minutes later, a lump of lifeless, burning flesh, and there could be no turning back. In many ways Karl's mortal experience was much alike to the bloody change humanity experienced when it entered its most profound change of phase, that being its transcorporeal transition that occurred a little more than a century later. Thanks to humanity's innate competitiveness, and the attendant exponential rate at which it developed its sciences and technologies, humanity ripped itself a new one. Except for a few who understood the meaning of exponential technological growth, humanity, like Karl, was caught by surprise. The manifold warnings of technology's encroachment did not help. Like Karl ignoring the encroachment of the American air base, twenty-first century humans, busy, tired, worried about the exigencies of their own lives, pressed on obliviously. Even Steven Hawking, with his simplified penny example went unheeded. Collect a penny today he told people in a popular book of his, then collect two pennies tomorrow, and four pennies the next, and keep doubling your booty, and by the end of a thirty day month you the pauper, thanks to this exponential growth, will have become you the millionaire to the tune of $10,737,418.21, but you wouldn't think so after only ten days. At ten days you'd have a mere $10.54 of change in your pocket. Even at twenty days, you'd only have $10,485.76—a tidy sum to be sure, but nowhere near a million dollars. And that was the rub. By 2008, thanks to many Moore's-like 'laws', many households sported more transistors than brain cells, but the portent carried with this milestone passed by unnoticed. At humanity's twenty day point, computers, powerful as they were, were still relatively dumb, and robots were still more toy than tool. Humanity's and little Heinreich's march toward the inevitable continued unchecked.

Indeed, miles away in Zurich, at the instant Karl's body impaled itself into the stump, ripping off his head from the collar bone up, a young physicist by the name of Albert Einstein, sitting at his desk before a fire, sipping warm tea, penned his name on a sheet of paper, formally declining a joint offer from the University of Zurich and the Federal Polytechnic, the ETH. The young man felt pleased with his work on gravitational waves. He no longer thought of his estranged wife, Mileva, nor of the illegitimate daughter, Lieserl, they had given away for adoption. The young girl would be lost to history...lost, that is, until the very end of human history.

Notions of practical wormholes and time travel were to be decades and forever beyond this young man's imagination. Throughout his life, moreover, this young man Einstein would never come to know much of Charles Babbage, then already dead in his grave 47 years, and even less of Herman Hollerith, already consulting at the Computer Tabulating Recording Company, or International Business Machines, as it would become known in 1927. The exponential evolution revolution was well in place, and under great acceleration under the exigencies of the first global war.

HITLERJUGEND

The first of Hitler's manifold youth programs began in 1926. Among the multitudes of children whom were selected by several Nazi party review boards throughout the whole of Germany, the brightest and most talented *Hitlerjugend* were collected into specialized programs. Without exception, the parents of these selected children were proud of their talented babes, the more so after receiving shiny medals stamped with Hitler's very face in recognition of their parenting skills, and their vital contribution to the German fatherland. Many of these beaming parents, in fact, took care to wear their medals on a daily basis, and most especially on Sundays to the envy of their fellow church goers. Naturally, no children of Jewish ancestry were selected for any of these youth programs, at least not so intentionally.

By 1926, little Heinreich, Baron Karl von Onsager's son, had grown into a strapping lad nearing his thirteenth birthday. Along with his father's blond locks and piercing blue eyes, Heinreich had inherited his father's easygoing nature and wry sense of humor. He was one of the children, among a thousand others ranging in ages from twelve to sixteen, who had been selected by the Nazis to attend the newly founded *Technische Gymnasium Hitlerjugend*, or the Castle Academy as it would typically be called during its early days. The imposing castle of the academy, situated in the shadow of the snow-capped Alps, came replete with battlements, parapets, a moat and a working drawbridge. In this old Teutonic setting, the isolated academy was designed to accelerate the training of the most talented German children in the hard sciences, and hence was considered to be the most prestigious of the various *Hitlerjugend Gymnasiums*.

Though Heinreich, to be sure, was an intelligent young man, his academic record left much to be desired. This was so mostly because the boy refused to apply himself. Heinreich rather preferred soccer fields then being troubled over his lessons. A little influence from his mother Frieda, the Baroness von Onsager, however, made all the necessary difference regarding Heinreich's prospects for admission into the Castle Academy. In the first place, it so happened that his mother owned the old castle at which the school was founded. In the second place, the baroness happened to be married to the Academy's first director, Wolfgang Boltzmann.

Director Wolfgang Boltzmann was the one and same man who had been Karl's faithful wingman during most of 1917, up through the day Karl died in late 1918. During this period the two young men, who shared similar backgrounds, had bonded as brothers. Naturally, after Karl's untimely death during the last day of the first world war, Wolfgang felt that it was his duty to do everything he could to help both Heinreich and Frieda cope with their loss. "I should not have let Karl fly home alone that day," he would often apologize to Frieda on his regular visits to her estate on the outskirts of Berlin. During these visits, Wolfgang would treat little Heinreich as if the boy were his own. Two years afterwards, Frieda, wanting Heinreich to have a father figure in his life, accepted Wolfgang's proposal of marriage. The man, after all, was a well-spoken, well-mannered, solid featured, dark-haired man of means who would provide Heinreich a good example, and she knew that he genuinely loved her son.

When in 1924 Hitler—who was good friends with Wolfgang—told Wolfgang of his idea of building *Hitlerjugend Gymnasiums* across the land to build a new and greater Germany, Wolfgang jumped at the idea. It did not take him long to convince Frieda to donate the use of Karl's old childhood castle to Hitler's good cause. "I promise you dearest, that such an academy will be an ideal place to tame our Heinreich and give him direction," Wolfgang had insisted with his arms wrapped around Frieda the day she finally relented.

"This Castle Academy notion," Frieda had said while beaming her famous smile, "is the first idea that you and that peculiar, little man Adolf have come up with that I actually approve of dear Wolfgang. I've thus decided that you may have the use of the castle on two conditions. That—"

Wolfgang, cutting Frieda off mid-sentence, smiled and kissed his wife several times on her neck and cheeks after her last, interrupted words. "Name them dearest!"

Frieda placed a finger on her husband's lips to assure his silence—he was trying to kiss her again—and then tapped out her conditions on the tip of his nose. "That, after I return from my tour of Greece with Baroness Zwiebach, we move into the summer house near the old castle so that I may be near Heinreich, and that I may be allowed to make suggestions regarding the curriculum."

A few months following Frieda's consent, it was there at the newly founded *Technische Gymnasium Hitlerjugend*, that young Heinreich—who had demanded on pain of death that his mother and stepfather keep his privileged identity a secret—started his life in earnest, for it was there that he met Lise Reber for the first time. The pretty, brown-haired girl with pensive brown eyes and a charming smile was a twelve—nearly thirteen—year old prodigy who swept the boy off his feet the very first time he laid eyes on her.

Like Heinreich, Lise and her family haled from one of the upper crust neighborhoods of Berlin. Happy and adventuresome by nature, the young lady had a warm and giving personality, and like Heinreich, she too appreciated wry, ironic humor. That Lise was part Jew through her mother's family had been, at this point in her life, lost to history. The fire which gutted the Registry of Birth in Bonn in early 1879 had destroyed the entire collection of birth records. Her mother's affluent family, being prescient about growing anti-Semitism—that term being first heard in Germany in 1879 as part of a description of anti-Jewish political campaigns in central Europe—swiftly converted to Catholicism, moved to Berlin and registered anew under proper German names. There, the family offspring, through prudent maneuvering, took care to marry among the finest German Berliners.

Belinda, Lise's mother, in fact, had married into one of Berlin's most celebrated medical families, the much vaunted Rebers. Studying music at university in the flower of her youth, Belinda, full of life and a bit uninhibited by old-world standards, had done her best to attract the attention of Max Reber. Max, following in his father's, grandfather's and great-grandfather's footsteps, was busy finishing his medical studies when he was smitten by Belinda. The feeling was mutual. From the start, Belinda had been drawn to Max's polite, conservative ways, his trim body, and his old man's pince-nez spectacles that he used to make himself look older. It didn't take Max long to notice Belinda's beauty and revel in her innate wildness. The hazel-eyed young lady smoked, drank and loved to dance to American Jazz deep into the wee hours of the morning, an activity Max found irresistible.

"We are so very proud of you Lise," Max and Belinda told their daughter at the Berlin train station as they waved their daughter off. Despite her overt genius and her maturity far beyond her years, little Lise still looked so young to her parents.

So too among the children chosen to attend the Castle Academy, again thanks to another unlikely accident of history, was Lieserl Einstein's frighteningly brilliant nine year old son, Hans Fritz. Of course, this lost lineage of Albert Einstein was not to be discovered until many decades later by a pair of determined scholars early in the twenty-third century using advanced forensic gene chip technology. Lieserl, it turned out, had been adopted by a pair of ardently religious Germans who had found her in a Hungarian orphanage in late 1904. Whilst touring the city of Pécs and its old Roman ruins, the Heisenberg's—who had remained childless for nearly ten years—stopped to pay their respects at the town's old cathedral and its centuries old orphanage. Amid all the crying eyes at the orphanage the couple saw that fateful day, Lieserl had been the only babe with dry eyes. Her plaintive puppy dog looks at once disarmed Herman Heisenberg and melted Eva Heisenberg's heart, so much so that by that afternoon, the couple started making arrangements to adopt the quiet toddler whom the monks described as pensive. When the Heisenberg's returned to their native Germany several weeks later, beaming joyously with their new daughter, they promptly registered Lieserl under a new name, Constance Heisenberg. Much to Herman's and Eva's delight, it didn't take long for little Constance to quickly attach herself to her favorite cousin, Werner Heisenberg. Werner, who was Constance's senior by a year, was already showing signs of brilliance by every measure. The lad, as it happened, would develop a talent for physics and eventually go on to win the Nobel Prize in that field in 1932. Playing with Werner formed some of the fondest memories of Constance's life. 14 years later, however, as a naïve fourteen year old girl, Constance bore Hans Frederic Fritz, and her life in Munich changed.

Constance, of course, hadn't meant to get pregnant at such a tender age, but she hadn't been able to resist the advances of Himmer Fritz, whose industrialist family employed her father as a senior technical consultant. Himmer, home from university for the summer, was in the flower of his youth and brimming with hormones when he first took note of Constance at a ball thrown to celebrate his family's latest factory opening in Bavaria. He was tall, blond, muscular and blue-eyed, exactly the kind of Aryan German the Nazis would later make posters of to depict their idyllic ubermen. When Himmer saw how much Constance had matured whilst he had been away, he became infatuated with the girl. It was Constance's long black hair and pensive eyes that drew him in at first that evening. Her bare shoulders and demure way of looking up at him on dance floor did him in.

From the start, Himmer wasted no time on his conquest, damned be it that Constance had just celebrated her fourteenth birthday. Obtaining her father's permission, Himmer invited Constance to go out rowing, then horse back riding a week later, then eventually out for several tours about the country in his other love, a new 1917 American Cadillac with its sparkling chrome embellishments. Without exerting any direct pressure, Himmer let Constance enjoy herself, relax, and get increasingly comfortable with his physical presence. When Himmer was finally certain Constance was attracted to him, he began talking to her about his feelings towards her—feelings even he believed in. "I love you," he would repeatedly tell her, "madly so to tell you the truth." Once this new phase of his began, it didn't take long for young Constance to succumb to Himmer's increasingly intimate advances. By the end of the summer, in fact, Himmer was having his way with Constance every day in nearly every way.

Two months after Himmer had returned to university, and his letters to the naïve girl were beginning dry up, Constance learned she was pregnant. To avoid tarnishing the great Fritz name, the senior Himmer—a devout Catholic—forced his wayward son to do the right thing, and to do so as soon as possible to protect Constance's family's reputation. Himmer and Constance were wed a mere week after the Heisenbergs had notified the old man of Constance's "delicate condition." From the time of his birth, life for Hans and Constance had not been very good under the Himmer household.

***

With a thousand lost looking children scurrying about the stone corridors of the old Teutonic castle, a certain amount of chaos was unavoidable during the first day of classes at the Castle Academy. With its parapets and towers wreaking of history, of bygone knights and their battles, the imposing place had for the most part been restored to its former glory. Nothing at all as Frieda insisted, never sparing any expense, was too good to house Germany's precious, best-of-the-best *Hitlerjugend* in an awe inspiring setting. Only a little final work—the addition a few electrical and telephone lines—was still to be done in the dungeons, where a nurse's first-aid station and several well equipped biology and chemistry laboratories were being completed, but this remaining work was not expected to delay Herr Boltzmann's tightly scheduled program.

Heinreich, carrying a handful of books, was headed towards his morning's first class when Lise, walking just ahead of him, being momentarily distracted by a huge oil painting of Adolf Hitler hanging in the corridor between two suits of armor, bumped into Hans, who stood quite a bit shorter than Lise.

The unexpected collision sent her books tumbling to the floor. One of the larger tomes, falling flat, set off several echoes in the long corridor. Heinreich, already well trained to act the noble blooded gentleman didn't waste any time in coming to Lise's aid. "Guten Morgen," he introduced himself while gathering the pretty girl's fallen books. "Mein name is Heinreich."

Lise smiled at Heinreich as she collected her books one by one from Heinreich repeating her thanks each time. "I am Lise," she introduced herself flashing a warm smile after taking the last of her fallen books. "Wasn't that some echo?" she asked whilst feigning an astonished expression.

"No kidding," the boy Hans piped in, interrupting Heinreich before he could respond to Lise's remark. "Based on the two reports of the echo," the diminutive boy continued, "I estimate this corridor to be of no less than one hundred meters in length."

Heinreich, transfixed for a moment by Hans' astonishing statement, quickly snickered with doubt before himself adding, "I surmise, additionally, that this old castle has a well equipped dungeon approximately ten meters beneath us." Of course Heinreich knew this.

Hans, staring at Heinreich with his mouth agape was visibly dumbfounded. "But how did you calculate that? It is impossible," Hans gibbered on.

Lise giggled. "He didn't dear boy," she interrupted, "Don't you see? He was merely joking."

Heinreich's eyes brightened. Lise clearly had sharp wits about her, and she was also quite pretty he noticed.

"Oh," Hans uttered a moment later still looking a bit flabbergasted, "I...I understand. My name is Hans Fritz and I am nine years old. My father often admonishes me for being too gullible."

Heinreich and Lise, both twelve years old and visibly larger sized then Hans, exchanged looks of surprise while Hans adjusted his round glasses. The minimum age for acceptance into the Academy was ten years. Hans, clearly, was something of an exception. "Well that explains your missing height," Heinreich declared as he reached out and patted Hans' shoulder. "By the way Herr Fritz," Heinreich began speaking to Hans as if he had known the boy all his life, "you most definitely remind me of mein dear cousin Walter who will be nine years old by week's end. He would have been here I suppose, but an unfortunate auto accident shortened one of his legs. I am most pleased to make your acquaintance."

While Heinreich reached out to pump Hans' hand, Lise, being mindful of the time, looked at her watch and noted that only two minutes remained before the start of classes. "Oh dear!" she exclaimed. "If we don't hurry we shall be late to our classes. Would either of you boys know which way it is to Professor Rudin's geometry class? I really don't wish to be late and attract any undo attention."

Hans, noticing that the number of lost-looking students walking about the dim corridors had greatly diminished, looked at his own over-sized watch in turn. "Nor do I," he replied, "and yes I do know where it is. If you will follow me, I happen to be headed that way."

Lise, turning to Heinreich with a questioning look, asked him if he would be joining Hans and herself at geometry class, or if, instead, he had to go to a different class.

Heinreich, acting as if he didn't have a care in the world, bowed his head ever so slightly in the way a baron's son should and replied, "but of course I shall join you, and I suggest that we run most urgently."

Despite their dash to class, Hans, Heinreich and Lise ended up being the last of the children to arrive at Professor Rudin's class, literally entering the room's threshold at the sound of the castle's grating claxon. All heads turned toward the trio while they gasped for air. Before them, professor Rudin, who sported an old suit, a long tuft of gray hair behind each ear and a pair of stern eyes, was sitting behind an enormous desk looking at his pocket watch. "How good of you to join us," he greeted the gasping trio with a hoarse voice. "Tardiness, dear children, will not be tolerated under any circumstances. Are we absolutely clear in this?" Both Lise and Hans, visibly embarrassed, apologized to the old man and quickly took a pair of empty seats at the rear of the class, but Heinreich remained put.

"Dear Professor Rudin," Heinreich addressed the old man from where he stood. "I should think you would be quite pleased that I arrived to class at all, especially if you knew of mein ordeal to get here."

All eyes, including those of Hans and Lise, now turned to Heinreich.

"Do continue," the old man replied studying Heinreich's confident expression.

"To begin with," Heinreich began, "before getting here, as you would agree, I had to walk halfway here." To add emphasis to his point, Heinreich walked halfway to an empty desk beside Lise. "Then half of halfway here." Heinreich now walked half the distance he had first walked towards Lise, who stared back at him with a frozen expression filled with consternation. "Then half again dear professor, and so on and so on. You do understand mein drift? I had to walk an infinitum of distances."

The old man straightening his aged tie stood up. He answered Heinreich's question with a simple nod and grunt in the affirmative.

"In fact, dear professor, I had to cover an infinity of halves of distances an infinity of times, but still, with the most assiduous determination I assure you, I managed to get here somehow. It's quiet paradoxical really."

Several of the students chuckled nervously after Heinreich finished his discourse. Studying Heinreich from across the classroom, the old professor began speaking after thrice clearing his throat. "The key to your paradox dear boy—what do you call yourself boy?"———"Heinreich sir, Heinreich von Onsager," Heinreich replied. The old man knew perfectly well who Heinreich was. Twenty-five years past, Professor Rudin had been the boy's father's personal tutor at the very same castle, and had been a dear friend of the von Onsager clan long before the deceased baron Karl von Onsager had been born. Moreover, Professor Rudin knew exactly what Heinreich was up to. The paradox that Heinreich had spouted out, as Frieda loved to tell her son, had driven his father, then a teenaged pupil, insane for weeks, until, at last, Professor Rudin took mercy on the young baron and gave him the critical hint, that being the issue of the partitioning of time. "The key to your paradox is that...," the old man continued with all eyes now turned to him—he paused to look around at all his pupils—

"does any one have a resolution?" he asked the children. Five seconds passed. "Anyone?"

Hans, with a sheepish expression, raised his arm. Professor Rudin nodded at the boy. "Well go ahead. Speak up."

Hans stood up. "The boy Heinreich can get to class," the diminutive child began, "because if he travels at no slower than a constant, finite speed, he will spend commensurately less time covering each halved distance—"

"Indeed!" Professor Rudin broke in cutting Hans off mid sentence. "This is indeed the resolution to one of Zeno's most perplexing paradoxes."

After a huff and twice more clearing his throat, the aged professor collected himself. With his face turned stern again, he turned his attention back to Heinreich. "In your case young lad, if needs be you should forget Zeno and run to my class even faster on winged feet." To everyone's surprise, a toothy smile then broke out across the old man's weathered face. "Now young sir, do take your seat beside your slow-footed friend Achilles and his procrastinating Polyxena." These names, drawn from antiquity, as Professor Rudin loved his Greek history, were clearly addressed to Hans and Lise. "In this class dear children," the professor continued, "we shall form groups of three, groups that will be selected by myself." Heinreich couldn't have been more pleased when Professor Rudin's first hand-picked group consisted of the none other than himself, Hans and Lise.

At the end of class, while the other children poured out into the corridor, Heinreich, knowing full well that the Academy was looking for team players, and would quickly eliminate any and all prima donnas, stopped both Hans the "genius boy" and Lise "the pretty girl" and lead the two of them into a quieter corner next to a water fountain. "I suggest that beyond our being grouped together in Professor Rudin's class, we should stick together like the three Musketeers and take care to look after ourselves."

Hans stared at Heinreich. "As in all for one and one for all?" Lise asked. "Exactly," Heinreich replied. "After all, as I'm certain is the case with yourselves, I am alone here."

Lise and Hans turned toward each other as if to ask themselves how about it. In each other's faces Lise and Hans read that they felt just as lost, displaced and alone as the other. Certainly at this point in time neither Lise nor Hans had any notion as to who exactly Heinreich was. Yet it was abundantly clear to both of them that Heinreich, obviously, didn't feel as alone or lost as they did, and based on his performance with Professor Rudin, the boy was obviously someone who was naturally well equipped to deal with the challenges—whatever those might be—of life at the Academy. Returning their gazes to Heinreich, the both of them simultaneously uttered in unison, "agreed." The trio exchanged quick handshakes before dashing off to their next classes, taking good care not to be late this time.

Later that first day, Heinreich did run into his stepfather, Director Boltzmann, and as agreed, both of them played out their individual roles. "Good day Herr Director," Heinreich said. "Good day young man," his stepfather replied barely raising his eyes from the roster he was reviewing.

***

Indeed, classes at the so-called Castle Academy began hard and fast. Being at the Academy, one of the more demanding gymnasium instructors would often repeat, is like trying to sip from a fire hydrant. Hitler wanted to build a new master society, and the children of the Academy were expected to become the avant-garde of that society. Accordingly, by Director Boltzmann's design, unrelenting pressure was put on the children to weed out riff raff.

"Are you alright Hans?" Heinreich asked his friend at the lunch meal formation—already it had been two weeks since Heinreich had met Hans. After suffering through two hours of difficult calisthenics, the nine year old looked beat and pallid, as is if he were near collapsing.

"Yes," Hans replied, separating the carrots out of his stew with his fork. "I just need to eat a little and I shall feel better."

Heinreich pat Hans on his shoulder while Hans chewed on a piece of leathery beef. "I have every confidence in you Hans." The din of voices in the busy mess hall, with spoons clattering, trays being emptied, students shuffling about for a place to sit, nearly drowned out Heinreich's words of encouragement.

When Lise joined the boys a few minutes afterwards, she took a seat opposite them on the long table. The first thing she did upon seeing the pallid state of Hans was to repeat Heinreich's line of questioning. "You look poorly Hans. Are you feeling well?"

Hans, used to this question from his mother throughout the years, shrugged his shoulders and took a bite of meat soaked in a dark brownish stew without lifting his eyes from the brown goop.

"It's not exactly what mother makes," Lise remarked looking her own bowl of stew.

"I hear it's made of horse," a neighboring, brown-haired boy interrupted. Upset that Hans had failed to block a vital pass on the soccer fields, the frustrated boy whose team had lost the match had intended to make Hans gag.

"It is a tasty horse if that is so," Hans replied without skipping a beat.

Heinreich chuckled. "That's the spirit lad," he added.

"Say, Lise," Heinreich continued now changing the topic, "how many girls have you lost from your section?"

Lise, wiping the corner of her mouth with the edge of her folded napkin replied that three girls had washed out of her section, and that a total of six had dropped out of her group so far.

"And one from my section and three from my group today alone," the neighboring kid with the brown hair butted in. The number of children eliminated became the talk of the table throughout the duration of the meal formation.

"Well," Heinreich concluded as he followed his last morsel with a gulp of water, "the three musketeers will prevail." Already each of the trio had covered the failings of the others more than once. Lise had helped Heinreich with a history assignment, Heinreich had helped Hans with pull-up exercises, and Hans had helped Heinreich with a math lesson. After two weeks under the pressure of the Academy's tough program, neither Hans nor Lise yet had any notion as to who Heinreich truly was, and just how much he would be there for them behind the scenes if needs be, but by this point in time Hans and Lise knew that Heinreich would be there them, and they in turn would be there for him. After growing closer to Lise and Hans, Heinreich did decide on one thing. He would not be shy of secretly using his mother to protect them should the need arise. He wrote her a thinly veiled letter to that end. *"Dear mother, I have quickly come to consider Hans as the brilliant little brother I never had, and Lise, who obviously has everything it takes to make it through the Academy, has already saved me from certain doom several times."* What Heinreich didn't write to the baroness was that he felt drawn to Lise and her pretty brown eyes in ways he couldn't quite yet understand.

Within two weeks, fully one tenth of the initial student body had washed out of the Academy's intense program. Many of the eliminated children—who had sobbed for home every night—had quit because they hadn't been able to handle the strain of being torn away from their homes. By the end of the third week, nearly another eighty children were attrited. By the end of the fourth week, however, the pace of children either quitting or being eliminated from the program had finally slowed by a significant fraction. Only twenty-nine had dropped out that week, and most of these eliminations had more to do with strain and fatigue than outright homesickness. By the end of the fifth week, the rate of attrition had dropped down to a trickle. Only nine students had dropped out, and of the nine eliminated students, six had been eliminated due to illness or injuries beyond their control. These six would be allowed to return to the Castle Academy the following year. By the end of the fifth week thus, things at the Castle Academy had pretty much stabilized. There were no eliminations that week. This is not to imply that the process of eliminating unfit students from the rosters ever had ended. It would go on until graduation day itself.

Injuries such as broken bones continued to claim a few more students at irregular intervals as the weeks turned into months. As before, these children would be allowed to return the following year. The few who were caught cheating on exams, however, were cashiered with great shame and, with their files locked away in Nazi party archives, would never be allowed to amount to much in Hitler's Germany. Lastly, by the end of three months, those children who proved themselves to be too arrogant and individualistic, were also identified and washed out on. The creators of the Castle Academy's programs had, from the outset, designed the pace and breadth of activities such that no one lone individual had a chance of making it through the rigorous demands without assistance from others.

The secret to success at the Castle Academy was teamwork, and it was teamwork above all else that was highly encouraged. Boys and girls, at first by random assignment, were thus banded together and encouraged to form strong bonds. Very soon afterwards, however, larger self-made groups, cliques and teams began to coalesce from out of the dormitories, classes and exercise fields. Hans, Lise and Heinreich, complementing each other's strengths, formed one of the better cliques, if not one of the larger groups.

Given Hans' diminutive size, it came as no great surprise that the rigorous calisthenics program of the Academy turned out to be his bane. He despised the exercise fields. Allowances, of course, were made for his age. Heinreich, on the other hand, was a strong, natural athlete who would complete his routines with an air of ease. He was fast, coordinated and motivated, but more importantly, he not only excelled for himself, as his instructors noticed, he also took care to help motivate and push the other boys to attain and exceed their own limits, especially Hans, who did make progress, even if only slowly. It was obvious to those in-the-know, and who had known the Baron Karl von Onsager that Heinreich was his father's son, and would one day become a great leader. As for Lise, separated with the girls on a separate exercise field, she was neither one of the strongest nor fastest. She was, however, never far behind the leaders of the pack. She preferred to pace herself and learn from the mistakes of the leaders, one of whom, being too aggressive on a tricky balance beam exercise, broke a leg and a collar bone, and was sent home in bitter tears.

In academics, Hans found himself in his natural element.

Lise too excelled, especially in physics, chemistry and maths. Heinreich, though he was by no means a dullard, achieved, as he always had, only barely average results in academics despite the constant urgings of Lise and Hans for him to work harder at his books. This was mostly so because the boy began spending far too much of his free time at the Castle Academy's glider club.

The Versailles treaty at the end of the first world war prevented Germany from building aircraft. To circumvent this ban, German aircraft manufacturers moved to other countries, such as Fokker to Holland. Domestically, in order to train pilots, glider clubs were thus encouraged all over Germany, of which the club at the Academy—supervised by the future commander of Hitler's Luftwaffe, Herman Göring—was one of the most well equipped and well located. Very few people at time—even at the highest levels in German government— knew about the secret bases in Italy and in the Soviet Union where German pilots had been conducting secret training exercises since 1925. Heinreich himself would not learn about these bases for some time to come.

It was on the glider fields where Heinreich most distinguished himself, and he did so very early on. By the end of his fifth month at the Castle Academy, Heinreich was the first student among the thirty-some members of the club to solo a glider off the nearby steep hills, soaring on that glorious day into the shadow of the nearby snow-capped German Alps. To everyone's amazement moreover, once Heinreich started flying alone on a regular basis, it didn't take him very long to master the various winds and thermals that swirled around the forested lands of the valley below, so that he never had to land on the strips that had been carved out of that valley. He would always make it back to the promontory from which he had departed.

"How do you manage to perform all those aerobatics and stay aloft so long?" an impressed Herman Göring asked Heinreich one day after the boy had returned from a particularly long flight.

"By watching the hawks soar, dive and kill their prey," Heinreich replied matter-of-factly.

"You do know son," the corpulent Göring continued, "I flew with your father during the great war. He was a great man."

Before replying, Hans took a moment to take in the vista. "Thank you sir. You would be happy to know that mein father wrote about you in his daily journal. One of your twenty-two kills, he wrote, saved his life. He was as proud of you as of every other pilot who earned the Iron Cross. I should add sir, that mein mother is very fond of your wife, the Baroness Karen von Fock-Kantzow." As the dinner hour was approaching, Heinreich excused himself from the future air marshal and walked off thinking about Lise, about how much he wished she could experience the beauty of soaring freely through air.

**

"Can you believe it Heinreich?" Lise asked Heinreich at that evening's dinner formation, "That we have been here six months to the day?"

Heinreich smiled. "My! I didn't realize we have been here so long." Heinreich was clearly in his element at the Castle Academy. "But then again," he continued, "look at Hans. Is he not an entirely different boy? After only six months, he no longer suffers those horrible asthmatic attacks, and is becoming a little force onto himself in team activities on the exercise fields. Just look how much taller he is. I suspect, the faster and stronger he gets, he will soon score his first goal on the fields."

Hans, busy chewing on a mouthful of the steamed carrots he had grown to love, could only return Heinreich's compliments with a sheepish smile.

"Now that Christmas is approaching," Heinreich started anew, "is there anything you miss in particular Lise?"

Lise considered the question for a few seconds. "My family above all else. And what about you Heinreich?"

"You," Heinreich let slip out, "and happy Hans of course," he added a second later with his face flushing. If Heinreich could have had it his way, he would have cancelled the Christmas break to have stayed near Lise.

The brief Christmas break of 1926 came and went. It was so brief in fact that when the children were home they hadn't enough time to readjust to old routines, instead feeling as if they were visitors in a familiar but yet strangely alien environment, visitors who didn't quite know how to act or what to do with their time. It was almost with relief when they returned to the routine of the old Castle Academy to commence their seventh month in the program. For those sturdy children who had survived to this point, life at the Academy had, for the most part, become a matter of comfortable habit. These children welcomed the arrival of 1927. In a few months Stalin would score a great victory over the Trotsky Leftists in Russia, the general labor strike that would start in May, and would last through the remainder of 1927, would soon begin to cripple England, and Hitler was about to publish his book Mein Kampf. Far removed, and yet at the very center of world events, the Castle Academy had been frigid and gloomy without its children, its fields of verdure having long since died. As for the children—like good little army ants living in a great big ant pile—the minute they returned and breathed life into the castle, they knew what they had to do, when to do it, and exactly how to do it. Regimentation into a collective had become a way of life.

***

"Thanks in great part to your help Heinreich, Hans has made great strides on the exercise fields despite his being the youngest member of the Academy," a calisthenics instructor named Wilhelm told Heinreich one sunny April afternoon while Heinreich helped the man collect all the soccer balls.

"I suppose Herr Wilhelm," Heinreich replied, "that far away from the noise, pollution and commotion of his native Munich, there could be no doubt that the open air of the surrounding forest, has had much to do with Hans' physical improvements."

Surrounded by rolling hills situated in the shadow of the snow-capped Alps, the air around the Academy was indeed pure and pristine. In the crowded city, Hans' eyes were always tearing and he found it difficult to breath, making him prefer—to his father's overt disappointment—libraries over soccer fields. Out and about in the mountain country air however, the boy, especially after his tenth birthday, began growing at a remarkable pace into a considerably larger and stronger boy. It seemed the more he ran about, the more his body responded to the new demands.

"I and others have noticed," Herr Wilhelm added, "that Hans is becoming a decent soccer player. Thus the embarrassment of being selected last for a given team is dying away for the boy. That dubious distinction is instead increasingly falling upon several of the other older, fatter boys."

Heinreich chuckled. "There is some justice to that Herr

Wilhelm. Those boys you refer to chided Hans very severely every time Hans tripped on the fields." Hans' personal outlook was indeed growing brighter by commensurate measures with every little victory he won on the exercise fields. In this way, the Castle Academy became his new home. As for Hans' real home, the more time he spent away from it, the more it became the last place he wanted anything to do with.

"Don't you miss your home?" Lise asked Hans one Sunday evening while the trio were opening care packages in Heinreich's dormitory.

"Nein. Despite my mother's best efforts, it is always too stodgy at my house. It is always filled with noxious cigar smoke and boring, ceaseless talk of business. On top of that, my father has never really liked me much. He wanted me to be strong and fit, not smart." This was, unfortunately, true.

Hans' father, Himmer Fritz, who had been compelled to marry the then nearly fifteen year old Constance shortly after he had been forced to return from university, came from a very well-to-do and highly connected family of industrialists. Himmer had fought against the forced marriage pleading that Constance's child be given up for adoption, but his conservative father, a devout Catholic who very much liked Constance and her similarly devout parents, and who wanted a grandson to boot, had insisted on the wedding. Otherwise, he let Himmer know in no uncertain terms, the reins of the family businesses would have been passed off to Himmer's younger brother Otto.

From the start, returning to university soon after his wedding, Himmer made himself as scarce as he could, claiming, if he were ever to graduate, that he needed time for his studies, even during the breaks. It thus came to pass that Hans spent his infant years learning to crawl and walk in his grandfather's country estate being none-the-wiser about the true nature of his parent's relationship and the ill feelings Himmer harbored for him. When Himmer finished his degree in electrical engineering two years afterwards, the "family" of three, at the insistence of Himmer's father, reunited. They selected a home in Munich near one of the family's largest factories. It was expected that Himmer would spend several years there learning the business side of things, as well as how to treat his "people." Himmer's father had espoused a philosophy, not untypical of his time, that leaders of industry, the one's worth their salt at least, had a paternalistic obligation to their employees as well as to the community at large. Excepting his senior managers, the old man like being called father Fritz by his employees.

Until his ascension to power over the whole of the family businesses, Himmer, who took the utmost care to hide his philandering in secretly rented apartments scattered about the city, had been a decent enough husband to Constance by the standards of the time. It wasn't too difficult to do so. Einstein's lost daughter, so young and naïve, had been easy to control and manipulate. The relationship between Himmer and Hans, however, never became more than tenuous at best. Himmer had wanted a strong, robust son like himself. Hans, who had nearly died at birth, had instead been born a diminutive and sickly child. Not the least source of tension between the two was that Himmer's career as a promising soccer player had been ended early in a tragic accident, and now his only son would surely not restore his prospects for vicarious glory.

Lise, having come from a beautiful home life, couldn't believe what Hans had said about his father's preference of brawn over brains. She paused digging through the contents of her parent's care package. "Why would you say such a silly thing about your father? I am certain he's very proud of you for being as intelligent as you are, like my father is of me."

"Because I disappointed, and I continue to disappoint my father," Hans replied with a sigh.

Heinreich, opening up a package of Italian pastries sent from Frieda, interrupted. "How could you have disappointed your father already? You just turned ten. It is absolutely ridiculous. Would you like some panettone? It's from mein favorite baker in Milan."

As the sugary smell of the Italian pastries wafted to Hans and Lise, their mouths watered. Heinreich handed each a chunk of the panettone.

"Well for starters," Hans continued, "I'll wager that neither of you knew that my father had been a famous soccer player. In fact, he nearly represented Germany in the 1912 Olympics as team captain. Unfortunately, a French player, or filthy swine as my father would have you believe, tripped over the ball and broke my father's right leg in two places at an exhibition match. It was a mere two weeks prior to the opening of the games. My father would thus not be playing in the 1912 Olympics, nor for that matter, though he received the best of medical treatment, would he be playing in the 1916 Olympics. His bone fused wrong, resulting in a shortened leg. For the rest of his life, my father must wear corrective shoes. Worse, he suffers great pain when he walks much further than a kilometer. Now, as both of you well know, I am a disaster at soccer and this upsets my father to no end. Do you understand Lise why my father prefers brawn over brains?"

Lise nodded no, but before she could utter a word, Heinreich, wiping crumbs from his lap, burst out laughing. "You mean you WERE a disaster, but look at you now! The fat

Steiner brothers are now the last selected for teams, and you are nearly two years younger than the next youngest boy."

Hans smiled when Heinreich reached out and messed up Hans' hair. "Yah, yah," Hans agreed and took another bite of his panettone.

As the three children proceeded to rifle through their packages, Lise suddenly squealed with joy. "Look boys, my father writes that he may be able to introduce me to Professor Schrödinger, with whom he attended university."

Heinreich looked dumbfounded, but Hans understood. "Heinreich, he is one of the physicists creating what is being called the new quantum mechanics."

At hearing this, the light went out of Heinreich's eyes. "I swear," he said looking at Lise, "if it weren't for Hans and I being around to distract you, you would spend all your spare time bouncing between the chemistry lab and the Academy's library." It was true. When Lise wasn't otherwise occupied with school business, or spending time with Hans and Heinreich, she busied herself trying to learn the matrix mathematics of the new quantum physics as it was being developed in several of the monthly scientific journals. An exciting revolution was once again stirring the airs of physics.

Indeed 1926 was a seminal year for the physicists on their march towards understanding and ultimately controlling the atom. That year they rid themselves of the patchwork of half-baked models that had been the old quantum mechanics and laid down the foundations to the new theory of quantum mechanics. Spooky action at a distance teleportation technologies and unimaginably powerful quantum computers would one day arise from this new theory. Lise somehow must have sensed this potential. While she taught herself to manipulate its abstruse mathematical constructs, she would feel as if she were on a direct path to touching the face of God. It became a kind of drug addiction for her, and her addiction shown through in her letters home, especially the ones to her father Max who, trained as a medical doctor, could appreciate only to a degree some of what Lise wrote him about.

Accordingly, Lise's erudite pursuit made Dr. Reber even that much more proud of his precocious daughter. Whenever he chitchatted with his patients, among the wealthiest and most influential Berliners, Lise was sure to be mentioned. Max, regarding education, had never made a distinction between male and female, and had raised his daughter with as much exposure to the sciences as his own father had done to him before, especially the more so after Lise, from around her third birthday onwards, became like a mop who lived to soak up knowledge with a great passion. By the time she turned six, Lise, in fact, would ask to spend her weekends playing with the microscopes and chemistry sets in her father's hospital's laboratory. And spending time in her father's hospital's laboratory during the upcoming summer break was exactly what she had planned to do until the possibility of learning directly from Schrödinger came about.

Heinreich, for his part, who was not in the habit of thinking things out days into the future, much less months away into the future, felt a tinge of jealousy. Hans, on the other hand, knew that he would have to suffer through his time back home, and he dreaded it.

***

New faces appeared when the fall session of 1927 reconvened. Lise, Hans and Heinreich—who had spent his summer in England with Frieda—had enjoyed their time away, but as the days of their summer break dwindled down, they had grown increasingly anxious to see each other anew, especially Heinreich for to see Lise. Other than having new classes to deal with, this time around the trio knew the ropes, and quickly settled into the Castle Academy's routine. It was like being home again. To wit, to Heinreich's relief, no one among the students was any the wiser regarding his true connections to the Castle Academy. The last thing he wanted was to be treated differently. He would have quit had that been the case.

If there was one thing the Academy did for Heinreich during the 1927-28 school year, it was to focus his mind. Perhaps it was because a part of him was chasing his father's ghost, or because it was in his genes, Heinreich had grown ever more devoted to flying. For the first time in his life, Heinreich began to define and chase his own dreams within a larger picture—which was exactly what the Castle Academy had been designed to do for children with talent in the sciences.

"I love flying," Heinreich would often repeat to anyone who cared to listen. If the weather happened to be beautiful on a given day, he would be sure to say, "what a great day to fly." Having mastered glider flying better than any of the Academy's instructors during the 1926-27 school year, Göring "promoted" the boy to instructor pilot early on during the 1927-28 school year. It was a great honor, but there was more to Heinreich's obsession with flying than merely flying. Growing more mature, he began to think of designing his own planes for a living—planes for transport, and, naturally, planes for racing. These "racers" he would sketch with great gusto, upon which, showing them to Hans, the young genius boy from Munich would mark with red ink, shooting down Heinreich's drawings for all their various engineering flaws. "Heinreich, you cannot have a canard so far forward without destroying the airplane's stability." Lise would be kinder. "It's a beautiful airplane, and if you would like, I, and Hans, if he wishes to, can teach you the mathematics you need to make more realistic sketches." After so many of these embarrassing exchanges, Heinreich finally relented to Lise's offers. After all, it would mean sitting side by side with the pretty girl that he was attracted to. Hans too, to a lesser extent, joined in the tutoring sessions. Heinreich, to his great annoyance, would always have to remind Hans to take it easy with him. "Remember Hans. I am thick headed. You must keep it simple or you will lose me."

At first, while Professor Rudin covered advanced algebra in Heinreich's maths class, Hans and Lise—who had been promoted to far more advanced classes—helped Heinreich work his way through rudimentary trigonometry and geometry a year ahead of the Academy's programmed schedule. That process would take the entire fall session as Heinreich was always a reluctant pupil. Only Lise's charms pulled him through his geometry and trigonometry lessons. After the Christmas break of 1927 however, Hans and Lise would manage to push Heinreich through the rudimentary calculus of one variable, and things mathematical would change in the lad's mind for ever after.

There was something useful to this calculus. That Heinreich could optimize functions that represented potential shapes for his airplanes with relatively simple operations was a real eye opener for the then fourteen year old boy, and suddenly Heinreich became hooked on mathematics with a vengeance. If a lesson could be connected to, say, finding the optimal angle of attack for a propeller, Heinreich would feel wowed. If no such application was evident at the outset of a new tutoring session, Heinreich would then spend the rest of his day trying to fish out some application.

Apart from looking for applications to aeronautics, Heinreich's time studying mathematics also provided him a means to spend time with Lise. He enjoyed feeling her warm body nuzzled beside his own in an isolated corner of the Academy's library, and the more he got to know the thirteen year old girl in the quietude of the library, the more his feelings for her grew. Lise, though Heinreich didn't say a thing to her during their first years at the academy, had a feeling that Heinreich had developed a crush on her almost from the start, and it gladdened her to think this. She too had her own feelings for Heinreich. She couldn't help it. No matter how tired she had ever seen him, or ill, or burdened down with homework, he was always filled with an underlying enthusiasm for life, and this drew her to him. Of course, she was well aware that she was not the only one who was ineluctably drawn to him. Heinreich's infectious personality drew in a whole mob of followers among many of the Academy's boys and girls. In a word, the lad was popular.

The time Heinreich spent learning mathematics was time well spent. Years later as a Nazi aviator, his training in mathematics would serve both him and his career in Hitler's Reich well.

Back at Heinreich's palatial home located only a few kilometers from the Academy, it went without saying that Frieda and Wolfgang felt extremely proud of their son's accomplishments over the period since classes had started in 1926. Being married to the Academy's director, Frieda knew full well through Wolfgang about the trio that her son had formed with Lise and Hans. If it weren't for Heinreich's insistence on anonymity, she would have loved to thank his two inseparable companions for having at last guided her son towards useful ends. In this regard, the Academy had been a smashing success, but with all of her husband's dinner table talk about the rise of the German nation against its oppressors and the need for Germany to grow its military forces—influenced in these thoughts as he was by his friend Hitler—she also couldn't help worrying about the fate of her son. The last thing she wanted was for another war to claim her son as war had claimed her beloved Karl, and these fears of hers would only grow after Hitler's occasional visits to her home.

"Sometimes when I hear you and Adolf speak," Frieda once told her husband in their boudoir after one such evening with Hitler, "I think the two of you would rather be off leading the children of the Academy to war. I don't like him in our house."

Wolfgang snickered. "I don't wish that at all. I don't, however, want neither Heinreich nor any of those children to live lives in an oppressed Germany. I assure you this is all that Adolf wants. You should be nice to the man. He is our guest, and you know he loves it here, near "his" prized gymnasium."

Frieda didn't say a word. The gymnasium was not Hitler's—not at all—it was hers.

The morning after that particular evening, Hitler took Frieda's hand after breakfast and forced her to look into his eyes. "I understand how you feel dear Frieda," he began telling her, "your husband Wolfgang tells me all. I myself have experienced the hell of war, and war is the last thing I want, but unless Germany rearms herself to dissuade aggressive foreign ambitions we shall suffer at the hands of the English and the French."

Frieda was embarrassed and disappointed with Wolfgang, who evidently couldn't keep his mouth shut.

"On behalf of all Germans, I wish to thank you baroness for your generous donations of land and cash. You are helping to restore German pride. I assure you that your first husband's death for the fatherland will not have been in vain."

Frieda, acting graciously, thanked Hitler for his flattery after he kissed her hand adieu. There was something eerie about the short man she didn't like. After Hitler departed, Frieda, more than ever, was stricken with a frightening presentiment that another great war would likely break out in the not-too-distant future.

***

After the Christmas break of 1927, the months of the spring session of 1928 seemed to fly by. When May turned into June, everyone's mind turned towards the impending summer break with anticipation. To everyone's dismay the rules changed abruptly. As opposed to the previous year's break which had started in early June, this year's break was to be delayed to late July in order to add more material. There had been recent developments in eugenics—the theory that the human gene pool needed to be cleansed of degenerates and sickly peoples alike through sterilization—which Hitler considered to be of the utmost importance, and wanted to disseminate to his pupils and faculty at the Academy. Hitler himself had insisted on the change. Needless to say, the announcement of the delay was not greeted with great enthusiasm and things—as the summer verdure bloomed—seemed to slow down to the point of pain.

"Can you believe this idiocy?" Lise asked Heinreich when she read the notice of the delay in the morning bulletin boards.

"Nein. So we are going to be here longer. How much longer?" Heinreich asked in his turn.

"Twenty-nine days," Hans replied.

"I will go surely mad," a nearby student complained. "My

mother will not be happy with this at all," the boy then added with a bitter tone on the edge of tears.

As the days of May dragged into June, many students, and even some faculty, couldn't help but to feel aggrieved thinking they should soon have been at home. When 14 June came and went, without exception even the most ardent students found it increasingly difficult not to think about returning home.

"I can't stand it anymore," Lise confessed to Hans one afternoon. "I can't wait to take our respite away from this gloomy, poorly lit castle with its dank walls. I feel so confined in them."

Hans, who didn't particularly like being home, agreed with Lise. "It can get gloomy here, and I hate listening to the echoes in the corridors while I try to sleep."

Even Heinreich nodded his head in agreement. "Just yesterday," he stated, "Helga Klein complained to me that she was going to miss her own birthday party thanks to the delayed year. The poor girl burst out crying into mein shoulder."

By the end of June, to Director Boltzmann's dismay, even the Academy's faculty found it difficult not to drift into daydreams of home and family vacations, and he told his wife and Professor Rudin so at a private dinner function at the nearby von Onsager summer residence. "It is summertime after all, and the weather is beautiful. Everyone at the Academy wants to be outside, and it is understandable," Wolfgang declared.

Frieda understood. "I can't blame them. Can you Professor Rudin?"

The old man smiled. "I suppose not. So why don't we teach our classes out-of-doors?" he suggested with a shrug. "I myself can present practical applications of mathematics. What better way to teach the tenants of spherical trigonometry than by building sundials and compasses out of sticks, estimate ranges, and even calculate the radius of the Earth during the impending eclipse? I'm sure your militant friend Herr Hitler would approve of this." In this, Professor Rudin was correct.

The old man's notion, Wolfgang concluded as his eyes lit up, had been a stroke of genius. Accordingly, within the week, Wolfgang encouraged the remainder of his faculty to follow Professor's Rudin's suit. Straight-away things began to liven at the Academy. Within a matter of days, the lethargy and torpor that had stricken both the student body and faculty alike, gave way to lively animation and curiosity.

To the students, the lectures being held out-of-doors, and the field trips they took throughout many of the nearby hiking trails were infinitely preferable to being cooped up within the gray and dank inertia of the castle's ever present darkness. "I don't think I could have stood another day cooped up in the old castle," Lise exclaimed to Professor Rudin during one of the first maths class field trips. "Me neither," Professor Rudin admitted to his star pupil.

By studying out-of-doors, the children learned their eugenics from the examples set by nature itself. "In nature, order is preferred," the children would be told with much aplomb by their chiseled face biology instructor Herr Grimmer. "The strong create the order and do not permit the weak to hinder them. Even an ant, when one of them falls, or becomes injured, becomes food for the remainder of the colony. Yet we humans foolishly defy this order by caring for those who suffer retardation, or those who, because of their limited mental capacities, resort to crime for their livelihoods. Should we not stop them from propagating these burdens on the rest of good society? Should we not end the miseries of their cruel existence? Would you like to be retarded and strain the lives of your parents for the rest of their lives? I think not."

To eliminate the weak made sense on the surface. In the

United States, after all, there was an enthusiastic American Eugenics Society who set up pavilions in state fairs to popularize eugenics as science fairs, and Herr Grimmer, like Hitler, followed the developments in America with great interest.

"Children, to judge for yourselves," Herr Grimmer once lectured, "you have only to consider the famous case in the American state of Virginia in 1924 against Carrie Buck and her mother Emma. They were, as the Americans called them, feebleminded and promiscuous, and justly sent to the Colony for Epileptics and Feeble Minded. Both women bore children out of wedlock, and Carrie's child Vivian was herself judged to be feebleminded at seven months of age. Vivian thus became a perfect candidate for sterilization, and their own Supreme Court concurred. Listen. I will read you their conclusion. "...that Carrie Buck is the probable potential parent of socially inadequate offspring, likewise afflicted, that she may be sexually sterilized without detriment to her general health and that her welfare and that of society will be promoted by her sterilization.""

That Carrie had been raped was not mentioned by the American court, and history would record that the sterilized and would-be degenerate Vivian would earn straight A's for deportment and make the honor roll from the Veneble School in Charlottesville in 1931. Only in 2001 would the Virginia General Assembly acknowledge that their sterilization law had been based on faulty science—at least the Virginians weren't as tardy with Carrie and Vivian as the Catholics had been with Galileo.

This kind of indoctrination in eugenics was certainly not restricted to the children of the Castle Academy. As in parts of America, the populace at large was itself steeped in this horrible theory through many an "educational" film, making it easy, not too many years afterwards, for the Germans to sterilize and even to kill off those of their citizenry afflicted with Down's syndrome, dwarfism, blindness, deafness, deformities, epilepsy, hereditary alcoholism and so on to the tune of 410,600 sterilizations and between 50,000 to 100,000 killed via merciful euthanasia. It was said of them—the so-called degenerates—that death was a kinder option than living out their lots in perpetual misery to the detriment of the greater society at large, and people bought this.

Only a few children at the Academy, and not necessarily the older teens, thought twice about the real and uglier implications behind their lessons in eugenics. Lise was one of them, though she kept her thoughts to herself. A few others, who had questioned the theory—because they had a sibling with Down's syndrome or the like living at home—ended up falling out of favor and were eventually dismissed from the Academy. Lise's silent "wisdom" about her younger cousin's Down syndrome had, in fact, been noted by several of the directors of the Academy. In actuality, it had been her mother Belinda, years back, who had warned Lise not to be forthcoming about her cousin Mark's condition. Belinda, recalling the lessons of her grandmother's tales, knew a thing or two about misguided fear and hatred.

When June finally turned into July, Lise's desire to return home grew into a nearly insufferable anxiety. "A new baby sister was born to my family three days ago. I should have liked to have been home for the occasion."

Heinreich expressed his empathy. "I'm sorry you missed out on the birth of your new sister.

Hans too, after his growing success on the soccer fields, also found himself very much wanting to return to his home.

By the end of June, standing taller than some of the ten year old boys, he had begun scoring goals on a consistent basis on the soccer fields, thanks in great part to Heinreich's coaching and Hans' own fierce determination. "When I get home, I shall show my father how much I have grown, and I will take him to the parks nearby and show him that I too can play his beloved soccer."

Lise smiled at her friend. "Good for you Hans."

Though Heinreich himself—now looking more like a young man than a boy—was just as anxious to return to his real home in Berlin as anyone else, there was something, or someone rather, that he couldn't get from out of his mind. His feelings for Lise were tormenting him, and the thought of being away from her for a couple of months was crushing him. By this point in time, after two years of being around Lise, Heinreich's crush on the girl had grown unbearably strong, and he was bursting to tell her so. He, being typically male, had no suspicion that Lise herself had growing feelings of her own for him.

Often, when Heinreich should have been paying attention in class, he would instead slip into daydreams. He loved, in particular, to imagine himself the hero, and was often influenced in his daydreaming escapades by the stories people told about his father, as well as by Jules Verne's novel *Around the world in eighty days*. Typically, Heinreich would play the heroic pilot flying back during the days of the first world war. Often he would also imagine Lise to have been captured by a band of crude and crass Americans. For what reason and to what end she had been captured, Heinreich was still too young and innocent to supply beyond the most obvious reason to a boy of fourteen years, which was, chiefly, to nurse the wounded Yanks back to health. In his daydreams, he'd imagine her being mistreated, underfed, scolded, housed in cold quarters and so forth. These self-made images would anger him. After fighting through arduous combat missions in his imagination, he would return to his air base and interrogate any newly captured Americans to determine Lise's whereabouts. One day, his daydream would usually go, a kindness paid to a prisoner, a cigarette or a can of ham given in exchange for information, would finally pay off. Lise would be with the 75th Artillery Brigade or the 22nd Infantry Division, or some such invented unit of the American military. He'd dash back to base, track down the intelligence officer, and find out where exactly Lise's captors where holding her. Determined to get his girl back, he'd fly away from his base without permission, and after a great and unlikely adventure, would rescue her from her ruffian captors.

Inevitably, Heinreich's daydreams eventually lead the boy to pursue a very stupid idea. Before the end of the spring session, he decided that he would take Lise, and Hans if he wanted to, on a glider flight damned be the torpedoes. The school year would not end before Lise had tasted the wonders of flight.

***

Sundays were the only lazy days at the Academy, and it was on a warm Sunday afternoon when Heinreich, Lise and Hans snuck into the glider yard. The impending summer break period was now but two weeks away, and everyone was filled with a giddy sense of carefree ease. It was a perfect time for Heinreich to carry out his plan without attracting much attention. After two weeks of begging and beseeching both Lise and Hans to fly with him, despite Hans' persistent doubts about Heinreich's so-called operation, both Hans and Lise had finally relented. Lise had only held back to torment Heinreich. In reality, she couldn't wait to get into the air. This was not the case with Hans, who was nervous about the idea of flinging himself over the edge of a mountain.

Even up to the last minute—with the three of them packing a picnic basket in Lise's room—Hans continued to express his doubts with diffident nervousness. "But Heinreich, don't you think you could get us expelled?"

To which Heinreich, looking amused, replied rather flatly, "I really doubt it Hans. They wouldn't expel the son of the Baron Karl von Onsager for flying. That would be utterly ridiculous!"

Neither Hans nor Lise still had any notion as to the true familial nature of Heinreich's connections to the Castle Academy. Hans, nevertheless, was taken aback by Heinreich's bold reply. "I guess so," he dissented a moment later not exactly sure how to address Heinreich's strangely blunt answer. "But I still don't like the idea. My father would not be at all happy to hear about this."

Lise, taking Hans by the hand, her other hand holding the filled picnic basket, interrupted. "Come now Hans. It will be truly wonderful even if soaring is only half of what Heinreich says it is."

The gliders of the Academy were stowed away in small hangers by the edge of the green hills that rolled to the edge of the towering slopes of the nearby snow-capped Alps. They weren't locked, but Niebergall, the gruff, oil-stained maintenance man who practically lived in one of those hangers, often milled around the place even on Sundays always looking for excuses to fix things that weren't necessarily broken. The gliders, as he often liked to say, were his. The pilots, much to his resentment, only borrowed them for an hour or so at a time.

When Heinreich, Hans and Lise reached the first clump of hangers about twenty minutes later—the hangers being located about two kilometers distant and two-hundred meters above the castle grounds—Heinreich stopped them short. "Stay here," he told them, and then, motioning his friends to stay put with his hand, he walked off towards Niebergall's office-hanger as stealthily as he could. When he reached the smallish wooden structure, Heinreich, raising his hand once again, stopped abruptly. "Quiet now," he whispered back to Hans and Lise. "Stay put while I go around and peak inside."

Lise smiled in anticipation as Heinreich disappeared around the corner, while Hans, with nervous movements, adjusted his glasses several times. While he waited to hear from Heinreich, Hans cast several furtive looks out and about to make sure they were not being watched. Heinreich, meanwhile, started advancing along the hanger's wall step by step. Once he reached the corner, he paused again to listen for a few seconds while Lise and Hans, standing about five meters behind, tensed up. When Heinreich finally poked his head around the entrance Hans swallowed hard. Suddenly Heinreich yelped out, "Herr Niebergall! How good to see you." Heinreich's unexpected outburst from around the corner startled Lise and Hans, and confirmed their worst fears—Heinreich had been caught—but the thought of running and hiding never struck either Lise or Hans.

A few seconds of silence passed with Heinreich out of sight around the hanger's corner. Transfixed, both Lise and Hans remained put where they stood fully expecting Herr Niebergall's inevitable appearance with Heinreich in tow. They had no doubt the three of them would soon be escorted back to the Academy's *charge d'affair* to face a heap of trouble for transgressing into an unauthorized area.

In reality, however, there was no Niebergall to be found. The place was empty. Heinreich, knowing full well Niebergall would be away that weekend, was simply playing out a terrible prank. As more seconds passed with him crouched around the corner, Hans and Lise, began to wonder what was going on. They exchanged quizzical glances and shrugged their shoulders. That was when Heinreich finally stuck his head back around the corner. Wearing a grin from ear to ear, he burst out laughing. "I got the both of you," he belched out unable to contain his laughter. "You should have seen yourselves."

Lise, who quickly understood the situation joined Heinreich in laughter. "You are a veritable hellion Heinreich," she retorted while Hans, looking pale-faced, let out a sigh of relief before sucking in more air. "I nearly peed," he declared a moment later with a second sigh.

"Come on now!" Heinreich finally beckoned Hans and Lise after catching his breath, "mein favorite glider is the red one in the hanger over there, the one nearest the promontory we use for launching."

Hans, looking at the indicated hanger added, "Of course it would have to be the red one."

Lise, becoming visibly excited at the sight of the red glider, again took Hans by the hand and urged him on. "Indeed it must be red Hans. Don't you know red is my favorite color. Now do come on!"

While Heinreich prepared the glider, Lise admired the sky. It was free of clouds and quite blue. Hans, on the other hand, walked over to the edge of the promontory and began surveying the forested plains of the valley stretching out to the east nearly 400 meters below. About two kilometers away and much lower below, he noted that several clearings had been etched out of the trees to serve as landing strips. Heinreich hardly ever used those landing strips down below as he could usually return to the promontory by riding on thermals and eddies of airs only he, it seemed, could see.

"This soaring stuff," Hans murmured, "is not for me." Just looking down the steep drop-off made him queasy.

Lise, standing beside Hans with her brown hair catching the breeze, didn't agree. "To soar like a bird. You'll have to try it Hans," but Hans nodded no with more emphasis than before.

"I'm ready," Heinreich called out a few moments later. "Hans, come help me pull the glider. I need it to be about 10 meters from the edge of the launching ramp where the wheeled carrier is. Do you see it?" The wheeled carrier was, essentially, a bucket with a stone weight that, when hooked to a glider, would pull it down a ramp to the edge of the promontory. After each launch, it would be reset with ratcheted gears, locked, and attached to a new glider.

As Lise approached the beautiful glider, Heinreich instructed her on what exactly to do. "Lise, after I climb in and set the controls, I want you to sit on mein lap. I will then strap us in and close the canopy. When I am ready, I will ask Hans to release the glider. Once it starts rolling on its own down the ramp, you will feel some bumps, but then we shall be off into the air before you know it."

Lise obliged, but Hans, standing by the release, hesitated again. "But won't the two of you be too heavy?" he protested as if the thought had just occurred to him.

"Nonsense," Heinreich retorted, "you should see fat Göring fly this thing. The man easily weighs four times mein weight. Of course though," Heinreich snorted with touch of disdain, "he never makes it back to the promontory, but that shan't be a problem for us. The winds are too favorable today for that contingency."

With Lise sat down upon Heinreich's lap and strapped inside the small semi-open cockpit, it didn't take Heinreich but a brief moment to get the glider's trim controls set for departure. "Okay!" Heinreich yelled out to Hans, "release us." Hans complied and the red glider began its roll down the slope of the launching ramp leading to the promontory's edge. A few seconds later the glider slipped off the edge of the promontory and Lise gasped as the glider's nose dipped down towards the rocky grounds far below. Heinreich was allowing the machine to build some speed before raising its nose. A moment later he caught an updraft gust of wind and off into the blue skies he and Lise went.

The vista from the air, as the glider climbed into the sky, was truly breathtaking. For nearly a whole minute, while Heinreich piloted the craft to ever higher altitudes riding atop the rising thermal, Lise could do no more than hold her breath and gape at the world shrinking beneath her feet with wide-eyed wonderment. "I always wanted to see the world the way the birds do," she finally managed to say after catching her breath. "It is more wonderful up here than I could ever have imagined it to be. I think now, dear Heinreich, I understand why you are so taken in by flying. Thank you."

Heinreich, enjoying the scent of Lise's hair and the warm weight of her body pressed against him in the small confines of the glider, smiled. Trying to fly the glider as smoothly as he could, he was obviously very pleased with himself. Nature, fortunately, was cooperating with him—the turbulence was nil.

"Look over there Lise!" Heinreich exclaimed, pointing out a spot down below. "Do you see Hans?" Down below on the promontory, at the location where Heinreich was pointing to, Hans, looking no larger than an ant, was running about waving his hands at them.

Lise gleefully returned Hans' salute. That was when Heinreich spotted a familiar hawk soaring nearby. "Look Lise, over there. Do you see the hawk?" Heinreich asked pointing the bird out. "It has its wings spread, and if I'm right, it will soon soar towards the top of the adjacent hill." At that instant the bird's wings caught the rays of the sun.

"I see it!" Lise replied excitedly. "It's so beautiful and graceful," she added. "Could we follow it?"

With a smile, Heinreich told Lise he would do his best not to lose the magnificent animal. He banked the glider towards the bird of prey, and just as he had predicted, the hawk rolled its wings and flew off towards the other nearby hill. To their great delight, Heinreich and Lise discovered that the hawk had built a nest there at the other hill, nestled atop a tall pine. On their first pass over the nest, they caught sight of a couple of hungry young hatchlings, and after a second pass, decided not to trouble the little family anymore.

A half hour of soaring in the silent skies passed by with Heinreich and Lise admiring the sights of the forested valley and the surrounding hills down below, as well as the imposing silhouettes of the snow-capped Alps coruscating in the rays of the summer sun.

Heinreich, gradually feeling at the same time both more emboldened and at ease with himself, considered telling Lise that he was attracted to her. He decided, though, to go about it in a roundabout way. "I wish," he began, "that mein father would not have died, and I would have liked him to be proud of me for trying to follow in his footsteps."

Lise, with that way of looking deeply into people with

her pensive eyes, turned back to face Heinreich and assured him that the baron would indeed have been very proud of his son. Before Heinreich could continue saying what he really wanted to say, Lise confessed some of her own inner feelings. "You know Heinreich, what I most want to do is to travel the world. I've wanted to do that ever since I could remember, ever since I realized the world was a round place filled with all types of strange places and even stranger peoples. I have my uncle Günther to thank for this burning desire. He gave me a colorful globe for my third birthday."

Heinreich, in his turn, confessed that traveling the world was what his restless mother Frieda did best. "After mein father Karl died, mein mother could never stay put, leaving me behind to spend weeks, sometimes months at a time, with relatives. At other times, while she cavorted around the world, she would leave me with friends of the family. She has continued doing this even after she married mein stepfather. As a young boy, I used to miss mein mother during mein forced sojourns in the houses of friendly strangers. I often used to wish mein mother would have been more a part of mein life." At least now, Heinreich thought to himself, married as his mother was to the Academy's director, Frieda had become much more a part of Heinreich's life.

What Heinreich didn't know however, but was beginning to suspect, was that the marriage between Wolfgang and his mother was much more a marriage of convenience. Frieda had married Wolfgang more because she had wanted her son to have a good role model than because she loved the man and his idealism. For Wolfgang, marrying the baroness was a way to repay the debt he felt to Karl, and also a way to advance his ambitions—not that he didn't love Heinreich. Wolfgang did take in Heinreich as if he were his own son, and this did draw

Frieda and Wolfgang together to love each other to a degree, but not enough to stop Frieda from traipsing about the planet. This habit of hers, however, gave Wolfgang the time he needed to move up within the Nazi's political hierarchies. Moreover, it was a fact that the Nazi's frowned on uncommitted men and women, hampering their careers to a great degree.

The conversation, intimate as it was between Lise and Heinreich, perhaps because they were more alone than they had ever been before floating in the open skies, flowed from one adolescent revelation to the next, continuing to meander between the two of them with an excited ease. Two hours in fact slipped them by unnoticed without Heinreich getting to say what he had intended to say, and it was good he thought that he had kept his peace and not spoiled the moment. He was getting to see Lise—who she really was when she was alone with him far away from other people—for the first time. Heinreich wished he could have carried on this way for hours more, but then he noted that the sun was sinking low behind a bank of clouds. The dinner hour as well, knowing that it would take time to get back to the Castle Academy after putting the glider away, was fast approaching.

***

Heinreich's original plan had been for him to fly both Lise and Hans for a just a few minutes, ten or fifteen at the most for each passenger, return the glider, enjoy the contents of their picnic basket and then walk about some of the nearby hiking trails, but it was too late for that. "I hope Hans won't be too angry with us Lise," he declared. "Do you realize we've been soaring for over two hours?" Lise, looking at her watch to confirm Heinreich's statement, couldn't believe how fast the time had flown. "We must return Lise," he told his happy passenger. "The dinner hour approaches and it will take us at least twenty minutes to return to the promontory, not to mention the time to put things away."

The thought of leaving the skies saddened Lise. "Must we?" she asked sounding genuinely disappointed. "I don't wish to land just yet."

Heinreich, sitting as he was behind Lise's direct sight, grinned. "Yes Lise. We must return. As you well know, we can't miss the dinner formation, but I promise I shall fly you again."

Had the young adolescents not shown up for the meal formation, their absence would have been duly noted, and a search would have been started minutes later. The Academy had its rules.

Reluctantly, Heinreich pointed the glider back to the promontory and began to lose altitude. "Do you see the leaves of the trees over there?" he asked Lise. "You can always tell the direction of the winds by the leaves. The bottoms of the leaves are more pale than the tops. When a breeze catches them, the bright, pale side of the tree points out the direction of the wind, that being towards the dark side of the tree."

This insightful observation pleased Lise. "You're an astute boy you know," she told Heinreich. "When you want to be, that is," she continued, lightly chiding him over his recalcitrance against listening to his professors and studying from—as he would put it—awful, dry books.

Heinreich rocked the red glider's wings back and forth to shake Lise about. "I know," he said. "When I want to be," he acknowledged.

The landing area was fast approaching when Heinreich and Lise noted that Hans had fallen asleep beneath one of the nearby trees. At 50 meters out, Heinreich began to flair the nose of his glider. The speed bled off so that by the time the glider touched down, it came to a gentle stop. There wasn't even a thud.

After Lise had unbuckled herself, she took Heinreich's hand and alighted from the plane. After Heinreich climbed out and secured the control yoke, Lise, to his great surprise, suddenly gave him an unexpected kiss on the cheek. After that, she thanked him for the experience. "I loved it up there with—with you," she declared with smiling eyes. Lise herself had meant to tell Heinreich something more while they were soaring in clouds, but, like Heinreich, she had not been able to bring herself to it either. In fact, it was she who had kept the conversation meandering.

While Heinreich stood before Lise, peering into her eyes at a loss for what to do after her kiss, wondering if this was the moment to confess his feelings for her, a sudden sense of self-consciousness overwhelmed Lise. She blushed, and before Heinreich could utter a word, she dashed off to awaken Hans. As Lise dashed off, Heinreich himself, following her with his eyes, felt flushed. "One day I will marry that girl," he swore to himself beneath his breath. For the rest of his life—even a century later—Heinreich remembered that day as one of the happiest days of his life.

While Lise busied herself trying to stir Hans from out of his slumber, Heinreich, pulling the red glider's towing handle, carefully dragged the long-winged machine back to its hanger. When Heinreich had at last restored things back to the way he had found them, and Lise and Hans were approaching the smallish hanger, a thick voice broke in from behind. "You three should take care not to miss the dinner formation." It was Göring's voice. "I hear," the fat man continued trying to sound stern, "they are serving pheasant today."

Caught by surprise, the three children grabbed each

other's hands and bolted off towards the mess hall. Their giggles and guffaws could be heard almost all the way back to the castle.

"Sorry I didn't get a chance to fly you Hans," Heinreich apologized when the three of them reached the castle grounds quite out of breath.

"Don't worry Heinreich," Hans replied while still gasping for air. "What I want to do…is to fly to the moon one day."

Lise, herself breathing hard, told Hans that she thought he was mad. Between his own gasps, Hans reassured Lise that, according to the papers, an American scientist by the name of Goddard had that very March successfully launched a liquid powered rocket from a place called Roswell, New Mexico.

Heinreich laughed. "Isn't New Mexico located deep in the middle of nowhere, where the American Billy the Kid gunned down people in the desert?"

Hans shrugged his shoulders. "I don't know this, but Goddard's innovative liquid fueled rocket," Hans insisted, "will change the world." Indeed, Hans was proved correct. Professor Goddard and his work would indeed become another falling domino towards the grand, tortuous cascade between the Greek's Icarus and humanity's solar system-wide mindspace wars.

***

When Heinreich had turned fourteen, Lise had asked him if he was beginning to feel more like a man than a boy. Heinreich told her that he hadn't thought of it much. "I guess I will be shaving soon," he had then added while feeling his chin and feigning to use a razor. Lise had found his antics amusing.

The break of 1928 came too soon for Heinreich. Lise and he—after their flight together—were beginning to break new ground, sharing ever deeper thoughts and quiet moments about hidden places they had made their own. "I'm going to miss you," she told him. "Our break begins next week."

Heinreich nodded in agreement. "Me too," he said. "But the next time you see me, I shall be shaving my whiskers—all three of them—like a man!"

Though Lise and Heinreich missed each other that summer break of 1928, even more so than they had during the summer break of 1927, they appreciated their time away from the rigors of the Academy. Heinreich and his mother—Frieda was itching to leave Europe with her son, who was now turning into a young man—dashed off to America by way of fast steamer. The plan was to tour America from coast to coast by rail. Foremost in Frieda's mind was to show her son that Americans were not crass, evil people as Heinreich had a way of implying when he talked about his dead father. "They are supporting our economy." Heinreich had understood years back that America was not his enemy. The only American he hadn't forgiven was the nameless one who had killed his father. Foremost in Heinreich's mind however, much to Frieda's relief, was spending time in America's west.

The voyage by sea was uneventful, at least from Heinreich's point of view. During several moonless nights sailing out in the middle of the North Atlantic, he did enjoy seeing the stars of the Milky Way as he had never seen them before. On several occasions Heinreich spent time with the teenaged daughter of a rich New Yorker, but he otherwise kept pretty much to himself, avoiding the bustling, chattering, socializing in the drawing rooms of the ship's upper crust peoples, the kind of folk Frieda and the young, teenaged girl preferred.

What most captured Heinreich's imagination throughout the voyage was thinking about Amelia Earheart's celebrated solo flight across the very same Atlantic only a month before. The Atlantic was so vast, it strained his mind to think of a woman daring to defy it all on her own in a tiny aeroplane. If a woman can do it, then soon, he thought, people will be flying across the oceans on a regular basis. Chuck Elwood Yeager, America's future greatest aviator, living in the woods of West Virginia, couldn't give a damn about Amelia at the time. He was five years old.

Heinreich thought New York was okay. He had to admit that its high-rise buildings were impressive, and the wealth of Wall Street impressed him. Frieda, on the other hand, really enjoyed her time in the big, bustling city. She attended several Broadway plays, and was taken in by the play Machinal, in which the then unknown Clark Gable played a part. She thought the young man was attractive and arranged to meet him after the show. It was then off to Boston the next morning. For the teenaged boy, Boston, and later Philadelphia, were in their turn as okay in the same bland way as New York had been. After all, he had seen plenty of museums in Europe, and America, by comparison was a new babe on the world scene.

Chicago, on the other hand, had some appeal to the boy. In February of that year Al Capone's gang perpetrated the St. Valentine's day massacre. Six Capone gangsters, dressed as cops, gunned down seven of Bugs Moran's men, filling their guts and brains with hundreds of rounds of steaming lead. The massacre was still the talk of the town in late July, despite the fact that Capone, going under the name George Phillips, had quietly moved into a fifth floor suite of the luxurious Lexington. Heinreich and Frieda, by pure coincidence, were put in the sixth floor and actually ran into the scar-faced "George" a couple of times in the lift. "Lise will be excited to hear I met the great Al Capone," Heinreich told his mother. In later years Heinreich would remember the man as having been very courteous.

After a few days exploring the windy city, came the long ride to San Antonio, Texas. Since before they had left Berlin, Heinreich had chosen this particular destination. He had recently read the story of the Texican revolution and wanted to see the famous Alamo for himself. "Very well," Frieda had assented, "its off to see some cowboys."

This part of the trip both pleased and saddened Heinreich. "This is the part of the trip I wish Lise could have joined us on." After San Antonio, the von Onsager's itinerary called for stops at Santa Fe, New Mexico, the Grand Canyon in Arizona, then finally San Francisco, California before returning to New York via the Panama Canal. Then it would be back to Berlin via fast steamer.

After checking out of the Lexington, Heinreich and Frieda boarded a train. A day and a half after departing Chicago, and riding through vast plains, the city of St. Louis came and went in the dead of night. So too did Oklahoma City another day later. "Mein God!" Heinreich exclaimed at one point during the trip down south, this country never ends! It took us days to get to Chicago from New York, and now we've spent three days in these endless plains, and we are only just in Oklahoma. I heard the conductor say it will still be two more days until we get to the Texas Alamo, and all the people speak just English! Imagine trying to do that—speaking just one language across all these kilometers—back home. By now we could be hearing Spanish or Portuguese in the background, or swimming in the ocean. This country is big."

Frieda, reading a fashion magazine beside Heinreich, smirked. "And you don't know the half of it yet."

Heinreich burst out laughing a second later when he realized that his mother had been more than funny. She had been literal as well. "No mother, I suppose I don't know the half of this country yet, but I can't wait to tell Lise all about it."

Life didn't start for Heinreich in America until he reached Texas, or the land of the cowboy and cactus as he preferred to call it. The deeper the train rode into the Lone Star state, the more he saw herds of cattle and gaggles of cowboys roaming about the lands marching their cows towards the rail stations. Unfortunately, it also got hotter and hotter as the Texas plains gave way to the hill country. For Heinreich, impatient as he was, the miles towards San Antonio's Guadalupe River began winding down clankity-clank by clankity-clank, until at long last they arrived at the town's old train station. No sooner did the von Onsager's hit the sack that evening at an inn by the Alamo, with their bellies filled with rice, beans and enchiladas, did they fall fast asleep.

To Heinreich's disappointment the following morning, the city of San Antonio turned out to be too modern, too electrified, with too many cars and too many tall buildings to be the old American west he had thought it would be. As for the Alamo, it was tiny and hardly worth getting excited about. Frieda and Heinreich, according to his wishes, cut their stay in Texas short, and departed for Santa Fe the very next day.

Heinreich's enthusiasm perked up when their train entered New Mexico and stopped to take on passengers. A sign by the little wooden station claimed Billy the Kid was buried nearby. Heinreich bought a pairs of postcards there and mailed them back to Hans and Lise. *I'm in the land of Billy the Kid*, he wrote his friends. On another pair of postcards purchased in Santa Fe, he wrote, *desert plains painted in red hues, open skies, harsh mountains, enormous storm clouds flashing bolts of lightening out in the horizon, this is New Mexico.*

New Mexico gave way to Arizona. Then came Heinreich's favorite place of his whole trip to America, the Grand Canyon. "Mein God mother!" he exclaimed when he walked up to its edge. "Imagine what the Spanish explorers must have thought when they first gazed upon this chasm. There they were, plodding along on their horses, oblivious, then, suddenly, they found themselves at the edge of a split running through the world, with a river running wild and fast far down below."

Frieda couldn't agree more with her son's assessment. "I wish your father could have seen this," she let out without thinking. "I am certain he would wanted to fly the whole length of it."

Frieda's remark made Heinreich imagine himself riding in an old biplane piloted by his father along the length of the Grand Canyon.

After visiting the Grand Canyon in Arizona, Heinreich was so taken in by its primordial beauty, he made a mental note that one day he would bring Lise to witness it herself. This thought—which kept rattling around in his brain— made Heinreich feel a sense of happiness that Frieda couldn't help noticing. The warmth that radiated from her fourteen year old son every time he mentioned Lise's name was just one of the many clues Frieda had to work with, all of them leading her to the same conclusion—that her son, even if he didn't recognize it, or didn't wish to divulge it outright, had a crush on this girl Lise.

Once, after having spouted out Lise's name half a dozen times in a single sentence, Frieda asked her son point blank, "Heinreich, do you like this girl Lise?" Heinreich, however, suddenly feeling awkward and exposed, looked back at Frieda with a blank expression and shrugged his shoulders in reply.

It was true. Heinreich did have a crush on Lise.

Something wonderful—something that Heinreich couldn't quite define just yet—had indeed transpired between himself and Lise that day they had soared into the skies with her by his side. What had started off as an unexpected attraction the first day Heinreich met Lise years back, had progressed into deeper feelings. Later, he would call these feelings love, but at fourteen, he wasn't ready to admit that to himself, let alone to his mother. Instead, to turn the tables, he asked Frieda a direct question of his own. "Do you love Wolfgang mother?"

Heinreich's question caught Frieda completely off guard. She hesitated for a moment, but then decided to be honest. "I do love him in my own way." She then quickly added, "I know that he is not your father, but he is a good man, and he does care about your future."

Heinreich smiled. "Then will you ever have a child by him?" he asked, egging his mother.

"I…I don't know. Now why are you asking me so many ridiculous questions?" Listening to his mother bumble her words made Heinreich laugh, and that made Frieda burst out laughing as well. She understood that her son had things he wanted to keep to himself.

***

While Heinreich enjoyed his time in America, Lise enjoyed meeting her new sister Olga. Little Olga was barely six weeks old when Lise returned home. The miracle of her sister's life, in fact, amazed the teenaged girl beyond description. Every week baby Olga grew in some palpable way. Her eyes learned to focus. She began attempting to grab objects in her field of vision. Lise—making furious notes in her diary—could tell that connections in little Olga's brain were forming and reforming at a frenetic pace, trying one path first, then another path, and so on, until her brain would arrive at an optimal, or near optimal solution.

Beyond studying her new baby sister, Lise also entertained herself in her father's hospital's laboratory, as well as at the main public library near the center of Berlin when she wasn't listening to Schrödinger's lectures. Though there was much she liked about life at the Academy, Heinreich notwithstanding, it exacted a drag on her primary interest in quantum physics. The break gave her time to catch up on her reading, and time to work through ever more of the esoteric mathematics being developed in the latest physics literature. The work of the Englishman P. A. M. Dirac was her favorite. That he was an avowed atheist, made Lise think about her own beliefs for the first time.

In the quiet hours of the evening before dinner, Lise would take walks about her affluent neighborhood lined with rows of aspen and maple trees and think about Heinreich. Between his postcards from America, she found it difficult having to wait to listen to his excited stories. She knew he would have many stories to tell her, probably enough to last weeks into the fall session, and that she'd enjoy much laughter hearing them. In her turn she would then relate to him the richness of her own summer's experiences with Olga and the family trip to the south of France. With Hans she would talk physics and higher mathematics, and about her heady talks with Schrödinger.

Lise realized she was growing older and rapidly becoming a young lady. The changes in her body told her that, as well as the maturing nature of her attraction for Heinreich. She wondered if she would tell him how she felt about him. Why she hadn't so far was beyond her. She suspected that Heinreich had figured out her feelings for him years back, as much as he suspected that she too had figured out his feelings for her.

Yet both had felt more comfortable keeping their feelings to themselves. Perhaps this year, she thought, I won't be able to keep quiet anymore.

For all of Lise's epiphanies that summer concerning the miracle of creation, her first questioning of her faith in God, the development of quantum physics and her feelings for Heinreich, she could no more have fathomed the thought that life, no matter how much science progressed, would ever be quantified into so many genes, and treated as so much software within her own lifetime, than that the world itself would soon go topsy-turvy before her young eyes. Decades later, in fact, working as a senior scientist in a cold and forbidding land far away from her native Berlin, Lise would often remember the break of 1928, with her baby sister cooing and giggling in her crib, and Germany enjoying relative stability, as one of the most special years of her and her family's life. Many years later she would also go on to note 1928 as the beginning of her first awareness of a new kind of active, manmade evolution. That was the year, rummaging through several physics journals, that she learned of the invention of Dr. Edgar Lilienfeld of New York. He had invented the transistor and had filed a patent for his creation in 1926 after leaving Germany to avoid the increasing persecution of Jews. Someone in the review article, Lise read, had written that these devices might serve to build electronic brains. Indeed, machines built out of these transistors, starting with the Mark I Perceptron in 1960, would go on to repeat baby Olga's exercises of learning new skills by means of feedback processes. Not too many decades later, versus the millions of years that it took between the rise of homo habilis to the rise of homo sapiens, miniaturized spintronic wizards would ascend from the hulking, lumbering beasts that computers were in the 1960s. Whole colonies of human minds would find residence in these new spintronic, quantum computing brains.

***

Hans' summer break fared much better than he had feared. The first thing the ten year old boy did when he returned home was to run up to his father's expansive study and interrupt an important business meeting. "Father," Hans broke in while gasping for breath, "I want you to see me play soccer! I've gotten quite good at the game."

Himmer, accosted as he was in front of the butler and an important business acquaintance felt he had no choice—in order to keep up appearances—but to relent to his son's ridiculous wish with a feebly forced smile. *Decorum in business as in family*, was the man's motto. "Would this Saturday do my son?" he asked Hans while the butler refreshed his brandy.

The old, portly businessman sitting across Himmer's desk guffawed with a raucous voice. "Why," he broke in between a thick puff of cigar smoke, "the lad is a chip off the old block!"

Himmer assented with an even more forced smile which didn't at all fool Hans. "Indeed he is," Himmer replied while patting his son on the shoulder. "Now do go say hello to your mother. I've important matters to attend to."

Hans, ecstatic with Himmer's reply, and determined to get through to his father, lunged at Himmer and gave him a hug before dashing off tell his mother the good news.

Constance—who would never know herself as Lieserl Einstein—was very pleased to see her son returned home. When Hans found her out in the grounds busily directing the gardeners for the evening's soirée, he ran into her arms. Unlike Himmer, Constance returned Hans' enthusiastic affection.

"My," she said to him, "you've grown so much. Look at you! You've filled in."

Hans, feeling obviously pleased, beamed back happy eyes filled with excitement. "And I'm much faster and stronger too mother. In fact, I just asked father to come see me play soccer with the neighborhood kids this Saturday. I have to go and tell Arnold and cousin Axel to invite me in their usual morning game."

For a moment Constance was taken aback by the news, but didn't mull over her somber thoughts too long before congratulating her son with a false enthusiasm. "Very good for you my son." Himmer, she thought, will be so disappointed with poor Hans, and the whole summer break is at risk of being spoiled. Poor Hans will be devastated.

The anticipation of the dreaded event—both Constance and Himmer secretly dreaded it—made for a tense week of terse conversation over dinner. While Hans excitedly related his stories of life at the Academy, and how much he continued to admire both Lise and Heinreich, Constance and Himmer would quietly stir their teas. "Lise is almost caught up to me in tensor calculus mama, and Heinreich has helped me perfect my passing skills on the soccer fields papa!"

Things, however, fared much differently than Constance and Himmer anticipated that hot July Saturday. Hans, in fact, who had spent the entire week practicing among the neighborhood boys he had previously done his best to avoid, played his heart out. Much to Himmer's great surprise, the boy actually played wonderfully well. Hans even managed to score a brilliantly executed goal during the second half of the match, and set up two more goals with well played passes to other teammates. Incredulous, Himmer could only sit back, hands clasped to his face, and mumble, "magnificent Hans," several times over as the match played itself out. Needless to say, the spirit at dinner later that day was quite elevated.

"You did quite well my son when you passed the ball to Arnold." Hans could only smile as his father continued talking and gesticulating tactics. "You should however be more mindful of the players rushing at you from behind. You nearly had the ball stolen from you several times, and that would have been disastrous."

Hans, soaking up his father's excited words, assented. "I will father. I will. Thank you for your wonderful advice."

After the servants had retired to the kitchen and Hans had gone off to his room, Himmer, swallowing down the last drops of his port, excitedly remarked to Constance that Hans, though still rough around the edges, had shown great potential, and that he would take his son out every Friday and Sunday to show him a few of his college day pointers. "But," Constance insisted, "you must be careful with your leg." Himmer ignored the remark. "My son will learn from the very best," Himmer insisted. Constance smiled. It was rare to see Himmer so happy, and even more so, happy about his son.

After Himmer lit his cigar he brought Constance back down to Earth. "My sweetheart, do you think the lad will begin growing taller? He'll need to be at least 1.4 meters tall if he is to have any hope of making the Olympic team?"

Constance sighed. "But of course he will," she said trying to remain positive. "Haven't you noticed how much he has already grown? The Castle Academy is doing wonders for him."

Himmer, puffed on his cigar and thought about Constance's reply for a minute. "I suppose good air and good discipline never hurt a boy," he added with a gay smile, and just as suddenly, Himmer became as enthusiastic about the

Castle Academy and the Nazis that ran the place as anyone ever was. He had first seen the place as a way to rid himself of his fledgling son. Now, perhaps, he saw it as the place to "grow" his son. "It must be a wonderful place," he concluded. Constance, more concerned with Hans' academics, couldn't have, nevertheless, agreed more with her husband's assessment of the Academy's role in Hans' development.

Like Lise, Hans would also remember that summer break of 1928 for many years to come. Most especially among his memories would be the sight of Himmer and Constance standing beside each other cheering him on at the soccer fields during the Saturday matches. It was a sight Hans got used to very quickly, and it confirmed his belief that all the coldness in his family had been his fault. But, thanks in great part to Heinreich's help, he had managed to change this on the soccer fields. When Heinreich returned to the Castle Academy that fall, he was the happiest he had ever been in his life.

***

As the American economy went in 1929, so too went the German Weimar Republic's economy. With England and France still licking their wounds, America alone had propped up the republic with massive loans during the 1920s. Much to the ire of the Nazi party, which nearly went broke in 1928, the economies of both the United States and the Weimar Republic flourished during the majority of that year. The Castle Academy itself, without Frieda's cash, would have been forced to shut its doors that year.

On the surface, it seemed all was well with both nations, but then came the October crash of the American stock market. It was that event, more than any other, that marked the beginning of the end of the Weimar Republic and the rise of Hitler's floundering political party. America, desperate for cash, started reclaiming its loans to Germany within 90 days of its Wall Street crash.

Almost immediately afterwards, German industry began shutting down. German unemployment went from a reasonable figure of 650,000 in 1928 to 1,320,000 by 1929. By 1930 unemployment hit a staggering 3,000,000, and it more than doubled that swollen figure by 1933 to include nearly one in four working Germans. Accordingly, the Nazi party attained 107 seats in 1930 and then another 288 by 1933, making their party the largest and most powerful party in the Reichstag. From that point on, there would be no turning the Nazis back.

Of the three children, it was Hans' family that was hurt the most by the economic debacle that began to tear Germany's entire social underpinnings apart. Of course, being cloistered away at the Castle Academy during spring session of 1930, Hans had no notion as to the severity of things back home. It was Constance, of course, who kept her son in the dark, especially about Himmer's "breakdowns."

The Fritz' were a proud, affluent family who employed several thousand people in several modernized factories scattered throughout the southern part of the country. Producing all manner of heavy duty electrical power transmission equipment since the turn of the century, the family had been highly successful industrialists for nearly two hundred years in the metal forging and gun making businesses. Their company had electrified many of Germany's and Austria's cities. The Fritz' thus, were people who were used to success and respect, not failure, and to a great degree they deserved what they expected, for what they conscientiously produced was always of the highest quality.

Himmer Fritz, who had only taken the reins of the family business late in 1927, was eager to make a name for himself. Instead, less than three years later, much to his chagrin, the man found himself forced to shut down most of his factories and cut his workforce by two-thirds. It was a crushing blow for a psyche not well equipped to deal with disgrace and bad fortune.

When Hans returned from the Academy in 1930, the first thing Himmer did was to measure his son's height with a tape measure. "Come here boy. Stand straight! Are you standing straight?"

Hans, standing as tall as he could stretch himself could only utter, "yes father I am." He held his breath while his father took his measurements.

"Wonderful," Himmer concluded with disgust after tossing the tape measure aside. "You've grown a mere two centimeters. You just don't measure up!"

Himmer stormed out of the house slamming the door behind him. Hans ran to the door and opened it only to hear his father cursing. "Good-bye Olympics, that damned black-haired Jew-boy will never be my equal! If it weren't for that filthy Frenchman, I would have had my glory. Now that is all gone." Himmer prided himself on his ideal looking blue-eyed, blond-haired Aryan features as they were being depicted on posters by the rising Nazi party. Though he could have no suspicion that Constance was Albert Einstein's daughter, he couldn't help but see the Jewish features in his "ratty" son.

Hans, feeling crushed, ran upstairs. "Mother, what is wrong with father?" he asked Constance. She was standing at the window facing the carport. Hans couldn't see the tears rolling down her cheeks. Constance quickly wiped her face clean, put on a smile, and turned to face her son. She didn't fool her son.

"Are you well mother?"

Constance, unable to restrain herself, burst out in tears again. "Yes dear boy. Come here." Mother and son embraced.

"Then why the tears mother, and why father's anger? I thought he was pleased with me."

Constance sighed. "Because the business is in a bit of trouble you know. Since America's collapse, things have been spiraling downwards here too. We are having to close several of the factories. I hope, and your father and grandfather believe, that things will soon settle down. I'm not making excuses for your father, but you must understand how much it hurts him to put our people out of work. Please understand he is feeling quite bad about himself."

Hans, taking in the news about the family's state of affairs, assented. "Yes mother. I understand." He didn't mention to Constance Himmer's comment about his being a Jew boy. It was the first time Hans had heard any such racist remark, and it had stunned him.

Later that evening, Constance called her mother in Austria. "Never mind that Hans has made tremendous progress in the soccer fields of the Academy," Constance began, "all his father cares about is that the boy will never be tall enough to play at university and beyond. He is such a stupid man." Constance then listened to her mother's reply sighing. "You should leave him my dear," the old woman said, repeating her old and worn out suggestion.

Things grew steadily worse at the Fritz household that summer break when Himmer began attending the increasingly boisterous Nazi party rallies in Munich, where the party's organizational headquarters were located. On 10 August 1930, Rudolf Hess, the future Deputy Führer, fanned the flames by circling his Messerschmitt M-23 over a leftist workers' rally to drown out their loudspeakers. Reading about the incident in the morning paper of the following day, Himmer considered Hess' action so bold, he joined the Nazi the party that very same day.

The more people Himmer was forced to lay off as the summer weeks passed by, the more he encouraged his remaining workers, "for the sake of Germany and their jobs," to attend the Nazi rallies and support the Nazi party. As it did with Himmer, the Nazi's ideological message of "blood and soil"—the belief that the products of German soil grown by the noble peasants and eaten by the people created a pure German blood and thus a pure Volk—made an effective impact on the mindset of rural areas about Munich and southern Germany as a whole. Appeal to German nationalism, also played well with both farmers and the middle classes of nearby Bavaria, where the Fritz' other factories were located—and there too Heinreich urged his workers to support the Nazis, so much so in fact, he began to rise in the party ranks.

Among the Nazis, Himmer was the epitome of everything Aryan. At home however, he was anything but an uberman. As the German economy continued to collapse, and more of his employees had to be let go, the man began taking to the bottle, and taking his frustrations out on both Hans and Constance. Things worsened rapidly until one evening, stumbling into the house in a drunken stupor after Hans had retired to bed, Himmer beat Constance in a fit of rage. Constance had refused to make love to the man after he had stumbled into their boudoir wreaking of booze and cigarettes. "Nein, nein," she had insisted while he slobbered over her. Before she knew it, he punched her three times in the face, bruising her left eye, splitting her lips and spraining her neck. Himmer, spent, collapsed upon the floor and fell asleep leaving Constance to weep through the night alone in the large bed.

By morning, Himmer felt remorseful and ashamed. The sun was high and Constance had left. He gathered himself, bathed, shaved, and then went out looking for his wife. Avoiding the servants by feigning to cough and be ill, he found her in the library. Himmer walked up to his battered wife, dropped upon his knees, and shamelessly spilled out a stream of his most contrite tears. "Dear Constance, I want to make another son...a better son...believe me, I don't blame you for Hans," he said to her between his heavy sobs.

Constance, unable to believe the ugly words that she was hearing, remained quiet.

"I promise my dearest," Himmer continued, "to never lay a hand in anger upon you again." When he had finished speaking, Himmer bent towards his wife, his hair still wreaking of smoke and alcohol, and kissed his her on her bruised cheek. "Go to bed now dear—the servants shouldn't see you in your condition," was all he could think to say afterwards. "I will tell the servants that you tripped and go see to breakfast myself. Don't fret dear. I will also see to it that the servants have their orders. After that, I must go and announce more job losses at factory number twelve."

When Hans saw his mother's battered face later that morning, and learned what his father had done to Constance, he told his mother that he wanted to lash out at his father and beat him with an American baseball bat. Despite Constance begging her son to forgive his father, Hans, remembering his father's Jew boy comment, refused. "He has a lot on his mind," she said to him barely containing her own tears of anger while Hans embraced her. "He loves us both and promised he will never do it again. You must forgive him."

Hans, shaking his head in his mother's bosom told her

that he would do no such thing. "I can't mother. I can't" A part of the boy was glad the fall session of the Academy would be starting anew in a few weeks. Another part of him however, was truly scared for the well being of his mother. "Can you not leave for a while mother, at least till things get better with the factories?"

Constance shrugged her shoulders. "I don't know my son."

To be sure, Himmer didn't lay a hand on Constance, at least not until four months later when, in a drunken stupor once again, he entered the boudoir and questioned Constance's lost heritage, calling her an orphan mutt who tricked him into marriage. Yet again he accused "her" son—brown-eyed and brown-haired and too short—of having too many Jew-like features. Constance called him a shameless pig. That's when Himmer struck her hard with the back of his hand. The blow sent Constance reeling to the floor, where she curled herself into a ball expecting another blow. This pathetic sight only enraged Himmer more, to the point that he kicked the woman in the back while screaming at her. "Even Director Hess has noted this…this Jewish-like taint in your son, and it shames me! Hans is not good for me at party headquarters." Himmer then stormed out of the room.

Constance, it went without saying, considered divorcing her rapidly disintegrating husband, but she had Hans to consider. The boy—though his father refused to notice—was flourishing at the Castle Academy. "Take pity on your son," she would often beseech him. "Hans is and will be a good Nazi. He will make you proud. Despite his young age, he excels at everything at the Castle Academy, and this should make Herr Party Director Hess proud of your son." To these remarks Himmer would only reply, "to the devil with the boy."

Hans' performance at the Academy meant nothing to the man, not even the fact that Hans, who with special help arranged by Frieda and her husband, Director Boltzmann, was actually working thru levels of mathematics beyond the reach of even Lise and Professor Rudin. "The Jew boy will never by an Olympian, much less an Aryan. The only good thing about the place is that it keeps him out of Munich and out of sight. If you leave me, as you so like to threaten me of late, I will pull the rat from that precious Academy of his! Hess is much higher up in the Nazi party than that doppelgänger Director Boltzmann. Is that clear woman?" In this way, Constance was effectively trapped. She wouldn't have been if she had only known who exactly Hans' friend Heinreich was, and that his mother Frieda would have made sure Hans' career at the Academy could not have been touched by Himmer, not to mention that Director Boltzmann, a close friend of Hitler himself, was fond of Hans.

"No matter how many times you ask me to leave him mother," Constance would say to her mother over the telephone, "I cannot. I cannot for Hans' sake." Other than the hospitality of their home, Constance's parents could not possibly offer Constance and Hans the special opportunities he was getting at the Academy. Her dear cousin Werner—making great contributions to the burgeoning theories of nuclear physics and quantum mechanics—did offer to take the boy under his wing, which very much tempted Constance, but before the professor could take Hans in, Werner accepted a series of lecture tours to the United States, Japan and India. The whole situation for Constance was a shame, for she was still a young and attractive woman.

***

Lise's family on the other hand, though they too were

most definitely affected by Germany's runaway inflation and collapsing economy, continued to do relatively well through the crash of 1929. In 1930, her father's medical business—thanks to the many affluent Germans who developed all sorts of imaginary ailments brought on by anxiety and depression over their weakening finances—did a brisk business that year. There was also a new baby girl in the Rebers lives to help keep their minds off the gloominess of Germany's fiscal collapse. Baby Nanerl, who was to be the Rebers third and last daughter, centered the family with her deep brown eyes and curly locks, especially so for Lise.

Since her first break from the Academy in 1926, she had enjoyed helping her mother care for her then baby sister Olga. Now, once again at home on break, Lise found herself helping to care for another newborn sibling, and marveling at the mysteries of life all anew. If Olga had been devilishly playful as a newborn, Nanerl, by contrast, was deeply observant and intensely pensive, just as her namesake, Mozart's sister had been. That both girls had issued from the same parents only deepened Lise's sense of wonder.

The troubles of the Reber household didn't begin until late September, when Lise and the children of the Academy were once again preparing to leave their home for the fall semester. When Lise thought about leaving home—feeling at once sad to depart Berlin and exhilarated to return to her old friends—she found it difficult to imagine that the break of 1930 was already her fourth break. Olga, who would run around the house talking up a storm, was already two years old. Lise herself, though she wished it wasn't true just yet, was becoming a young lady with undeniable curves and bumps showing up and changing her figure. Five years, she realized, had brought her a lot of changes, but other than feeling this nervous reluctance about her maturing body, everything, both at home and at the Academy, seemed to be going well for her. Then seemingly from out of the blue, Lise took seriously ill.

At first, Lise's ordeal began with a mild fever a few weeks after her sixteenth birthday. Other than feeling weak, she was otherwise well. She didn't even lose her appetite at first, and her malady remained ordinary enough for several days. No one in the family thought much of it. She had the sniffles and that was all. Then much to her surprise and embarrassment, Lise got her first period.

The very next day Lise awoke feeling stiff and achy, and was running an alarmingly high fever. Throughout the day her fever worsened, and her legs, in particular, began to ache to the point of tears. By evening, Lise's turn for the worse continued unabated. The pain in her upper legs grew to such an extreme that no other pain killer except for morphine, which her father Max was reluctant to use out of fear of addicting his daughter to the drug, could assuage her misery. He gave her an injection of the potent pain killer only after Lise, who had always been a tough little girl, started begging to be knocked out. That was when Max, who was known for his extraordinary equanimity, realized something was truly wrong with his daughter, and fear gripped his gut.

While Lise's mother Belinda—still plump from having carried Nanerl—stayed by Lise's bedside that first night, washing her daughter's febrile body with cold compresses, stroking her hair, praying to God for mercy, Max himself dropped everything else on his plate and took personal charge of his daughter's medical care. To get to the bottom of things, he decided to have all manner of blood work conducted. He drew the samples himself that night and had the samples rushed off to the well equipped laboratory at the private hospital at which he practiced. To Lise, the drawing of her blood was a miserable, painful ordeal, but she knew it had to be done and she suffered the drawings with great courage. The technicians, whom Max had awakened by telephone in the dead of night, and whom were very loyal to him for the way he both treated them and paid them, rushed to their lab and got to work on Lise's sample the moment they were dropped off at two o'clock in the morning. Lise, meanwhile, perspiring, moaning and tossing about, eventually fell into a morphine-induced stupor

By noon of that same day, polio, much to Max's relief, had been positively ruled out. So too were ruled out typhoid and many of the other such diseases common in those times before general immunizations. Most of the other test results came back negative as well. A few of the tests however, much to Max's angst, came back with inconclusive results. That meant more damned poking and testing, and more pain for Lise to endure.

Another day of detective work followed while Lise continued worsening. Unfortunately, the new round of tests—which required growing cultures anew in specially treated agarose environments for several hours at a time—played themselves out in the laboratory once again to no avail. Without benefit of the gene chips that would pervade medicine less than a hundred years hence—chips that would generate nearly instantaneous results from a micro liter of blood—the diagnostic processes of his epoch made Max feel as if he and his technicians were groping in the dark.

The whole affair for Max, who was exhausted to the bone by this point in time, collapsed him into a disconcerted mess of wrecked nerves, especially so after he accidentally dropped a test tube filled with Lise's blood. Later that day, while offering his most emphatic apologies to his daughter back at home, the man was forced to draw yet another sample from Lise, who by this point was looking horribly haggard after having suffering through nearly three days of high fevers. She forgave him. "It's okay," she said to him with a weak smile. "I love you daddy." Max nearly lost his composure.

Whatever it was that ailed Lise, it remained a stubborn mystery that continued to beat her down the next, third day. Max, though he kept it to himself, began to fear the worst for his daughter. By the noon hour of the fourth day, Lise had lost nearly three kilos and had slipped into a dizzy delirium. Breaking out in a cold sweat, at times she would mumble Heinreich's name and rant about flying like a hawk for hours at a time. "Heinreich," she kept repeating, "why don't you answer me? Where are you dear boy. I want to go flying to the south pole. I am so hot. I want to see a glacier. Please answer..."

Heinreich—whom the Rebers knew well by this point in time—was summoned by telegraph that afternoon.

***

During, and ever since Lise's first summer break from the Academy in 1926, the Rebers had suspected that she had taken a fancy to Heinreich, for all that she had talked about to Max and Belinda that first summer break was about Heinreich. He was such a good glider pilot she would say, and he was so kind to Hans, and such a good athlete, and so on. Her hair-brained day of soaring into the skies at the end of her second year at the Academy, soaring "out among the hawks" with Heinreich, finally revealed to the Rebers exactly how much their daughter was taken in by the boy. That she had done such a stupid thing had scared her parents half to death, enough so in fact, that they felt compelled to make overtures to the Baroness von

Onsager at the end of her and Heinreich's trip to America. They wanted the personal assurance of the baroness that no such foolishness would happen again.

Like other affluent Berliners, the Rebers, at the time, knew hardly more of the Baroness von Onsager than of her reputation as a well-connected socialite. Many of Dr. Reber's patients were also acquainted with the lady to various degrees. Neither Belinda nor Max, however, had any notion that the baroness, other than being Heinreich's mother, had anything to do with the Castle Academy. When they drove to the Academy to ask for Frieda's telephone number concerning Heinreich's and Lise's flying incident that summer of 1928, they were surprised to be sent directly to Director Boltzmann's office. Wolfgang, who enjoyed meeting the parents of his students whenever the opportunity presented itself, welcomed the unsuspecting Rebers with open arms.

It was awkward for the Rebers when Director Boltzmann warmly greeted them himself and invited the couple into his ample office, filled as it was with shelves sporting an extensive library and all manner of artifacts from Africa. "Come in dear Dr. and Frau Reber. Please have a seat. To what do I owe this honor to?" Hesitating for a moment, Max cleared his throat. The last thing the Rebers wanted to do was to jeopardize Lise's standing at the Academy by upsetting one of the other parents—an important baroness no less—but in no uncertain terms did they want their daughter flying again with some cocky, inexperienced boy pilot.

"Dear director, we are here about our daughter Lise. We are concerned for her..."

Max paused not quite knowing how to proceed. Director Boltzmann broke in during Dr. Rebers' awkward moment. "Concerned about what? Lise is one of our best students, standing head and shoulders above her would-be peers." Director Boltzmann was clearly puzzled.

"We are concerned," Belinda started, "with her safety. She admitted to us that she was taken on a secret flight aboard a glider piloted by the son of Baroness von Onsager, a young lad by the name of Heinreich before the end of the spring session."

Now it was Director's Boltzmann turn to feel awkward. "Your daughter, she told you this herself?" Both Max and Belinda nodded in the affirmative. Director Boltzmann, turning red, called his secretary. "Frau Reichle, please contact my wife." The Rebers, after hearing the word wife, could only stare at each other fearing that they were about to hurt Lise's prospects at the Castle Academy beyond repair.

While Director Boltzmann waited for Frieda's call, he assured the nervous Rebers that he would get to the bottom of things and, to their relief, begged for their forgiveness as he explained who exactly Heinreich, the baroness his mother, and he himself was. A few minutes later, when Frieda called back, Director Boltzmann asked if Heinreich was around the house. He was. "Then dear, please let me speak with the boy."

Max and Belinda listened as Wolfgang continued, "Tell me Heinreich, did you fly Lise Reber in a glider before the beginning of the summer break?" Director Boltzmann turned pale when he heard his stepson's reply. "Pass me back to your mother...Frieda, would you have Berta prepare a dinner for five...Yes dear, I will explain later." After hanging up, Director Boltzmann stepped out to ask his secretary to schedule a meeting with Göring.

When Director Boltzmann returned, he had a look of shame and guilt about him. "I am so sorry. Your daughter's life should never have been put at peril in such a careless way. My stepson has admitted the veracity of Lise's claim. I assure you that it will not happen again. Please, I insist, join my family for dinner and allow us to make amends. My stepson will give you this assurance himself."

To their relief, the Baroness von Onsager received the nervous couple later that afternoon with the greatest of grace and cordiality. Attired as if planning for a night about town, she was as stunned to learn about her son's puerile prank as her husband had been. "Welcome Dr. and Frau Reber. Please come in." The little isolated summerhouse as Frieda called it—the oversized mansion really—astounded the Rebers. It was built like a fortress. Little wonder Hitler enjoyed spending time there by "his" Castle Academy.

That evening, to their continued surprise, the Rebers learned that the baroness had endowed the Academy with a land grant, the castle itself, and a generous donation of cash. Thanks to Frieda's chattiness, they also learned that Lise was considered to be one of the Academy's most distinguished and liked pupils, and that, together with the boy Hans, she and Heinreich were inseparable. Lastly, having by now guessed it, they learned of Heinreich's desire not to have his familial relationship with the Academy revealed. "We promise," the Rebers assured Frieda over coffee, "that your son's secret is safe with us. Not even Lise will know." The baroness, in her turn, assured the Rebers that evening that Heinreich would not be allowed to fly Lise again. "I offer you my deepest apologies concerning my boy's misdeed." Secretly, though she did make Heinreich personally apologize to the Rebers for his stunt before retiring to his room, Frieda was pleased with her son's moxie. "He's a handsome lad," Max noted afterwards. "And he is quite charming," Belinda added. "Little wonder that our dear daughter is so taken in by him."

Ironically, before the Rebers departed for Berlin, Heinreich managed to persuade Max to take a flight with him in a larger, two seat glider. After spending an hour soaring in the air, Max told Heinreich that it was the most exhilarating experience he had ever had. Heinreich bowed his head. He wanted to say that was Lise's sentiment precisely, but decided against it. "I'm pleased you enjoyed yourself Dr. Reber." Then a spark of inspiration struck the young man. "Dear Dr. Reber, if you wish, I would have no issue if you were to reveal my secret to Lise. Perhaps then she could visit?"

To both Heinreich's and Lise's mutual delight, that first meeting between Wolfgang, Frieda and the Rebers precipitated a new and warm friendship between the families. At the end of that year in fact, Frieda and Heinreich—at Lise's urging—celebrated Christmas at the Rebers' winter estate, where both families enjoyed a week of skiing. It went on from there. The next year in fact, during the summer break of 1929, Frieda took both Lise and Heinreich on a grand tour of Spain and North Africa. The then seventeen year old young man and sixteen year old young lady got to spend many a summer evening walking about private beaches talking about all things between heaven and earth. Towards the end of that same summer break, while Frieda pressed on to Italy, Heinreich spent a few weeks with the Rebers in Berlin because, as he claimed to his disbelieving mother, he had taken an interest in the study of bacteria and viruses. At Lise's urging, Dr. Reber had offered the boy an opportunity to work in the lab of several of his distinguished colleagues. The Germans Max Knott and Ernst Ruska, who were close acquaintances of Dr. Reber, were busy perfecting the world's first electron microscope, and Lise had wanted Heinreich to join her while she helped out in Knott's and Ernst's laboratory.

Even if they hadn't admitted it to each other, by their

fourth year at the Academy, it was obvious to everyone else that Lise and Heinreich had been suffering raging crushes for each other almost since the moment they had met. The only mystery that remained was when they would finally admit it to each other.

***

Needless to say, Heinreich, who had been touring the south of France with his mother—it would be his last summer break from the Academy as he was seventeen and was to graduate the following year—was devastated when he received the Rebers' frightening telegram about Lise's dire condition. With his mother's approbation, he quickly made hasty arrangements to return to Berlin as fast as he could. Later that day, while heavy rain clouds began to form over Marseille, Heinreich wired back the Rebers from the train station notice that he'd be back in Berlin within no more than two days. Thanks to the ongoing partial strike of French rail workers—and the train stations being filled with a tangled mess of tired, frustrated people—he could do no better than this. If I could only fly straight over this mess he thought. Heinreich, eating himself alive with fear, ended up having to wait nearly fifteen hours in the thickly congested station to board his train. It was by far the longest and most painful wait of his charmed, young life.

The heavy rains which began pouring at the noon hour only increased his morbid anxiety. The din of the crowd, keeping him from focusing on his thoughts, angered him. If he couldn't be by Lise's side, he at least wanted to be alone.

Once aboard the crowded train, Heinreich took a window seat near the rear and stared out into space as the train rolled out of the station and into the rainy, gray overcast. Passing an old cemetery on the outskirts of town didn't help Heinreich's dark mood. Looking at his watch in exasperation, he decided to ignore the passing scenery and instead tried to write into his diary. Lise had convinced him to keep a diary the year before, but Heinreich found it difficult to maintain the habit.

The hours blanketed beneath the gray skies and cold drizzle trickled by. Heinreich cursed under his breath at every stop. When the first meal service was announced, he pressed his face to the cold glass of his window and stared into the gray ignoring his hungry gut. On and off between the interruptions, the conductor punching tickets, the next meal service being announced, a baby bursting out in tears, Heinreich, enclosed in his own mind, slipped into and out of reveries of happier times. The rhythmic background of the train rolling over the rails eventually lulled him into a state of dull, anguished torpor.

Heinreich's spirits lifted a little when he remembered how happy, and how unexpectedly surprised he had felt the last time he had seen Lise only a few weeks before. It was on the last day of classes at the Academy when it happened. Lise and Hans, whose respective parents would pick them up later that breezy, sunny day, had helped Heinreich lug his over-stuffed baggage to the Academy's bus depot. From the Academy, the bus would take him directly to the train station, and from there he would be off to meet his mother in France. Even though he had turned seventeen months before, Frieda planned to celebrate his seventeenth birthday in France in a large chateau she had rented for the occasion. To Heinreich, it seemed that the break of 1930 was just going to be another break, and nothing more.

A sudden jolt as the train rounded a curve snapped Heinreich awake. For a second he let himself believe that the train had reached Berlin. It hadn't. A few minutes later, Heinreich, sitting once again with his head hung low, closed his heavy eyes and slipped away into a leery, somnambulated state.

Once again Heinreich began recalling his last day at the Academy. He smiled as the visions filled him. That last day it had been warm, and everyone at the Academy, he clearly remembered, had been cheery, cutting many a silly joke. The last thing he had expected was that his world would change that day, but just before he stepped aboard the bus, Lise had abruptly stopped him and placed a note in his hand. Then, in the style of the French, to Heinreich's great surprise, she embraced him and kissed him once on each of his cheeks. "I believe the Frenchies call this greeting the *bise*," Lise then exclaimed with a nervous smile still holding Heinreich's hand with the note.

Lise had struggled for many weeks with how exactly to express her feelings to Heinreich—who, standing transfixed by the running bus's open door, was feeling light headed and warm all over. Without reading her note, he had a good idea as to what it said by just reading the longing, nervous look in Lise's eyes. And he intuited why she had waited till he was about to enter the bus at the threshold of the summer break to reveal her feelings. It was the safest thing she could do to protect herself against the possibility of making of fool of herself, and simultaneously the most powerful way to make Heinreich yearn for her if, as she strongly suspected, he had feelings for her.

The gruff bus driver, his patience tried, broke the moment. "You won't be kissing too many of them pretty French girls if you don't climb aboard." Heinreich—who then nearly blushed when he realized that more than Lise's eyes were focused on him—boisterously agreed with the driver without ever removing his gaze from Lise. "I promise I shan't kiss *too* many of the pretty ones on any given day." The young men aboard the bus—to whom Heinreich's remark was addressed—laughed. Hans, standing beside Heinreich and Lise, then piped in with a perfect suggestion. "If I know my French customs, I believe you are obligated to return the gesture to the lady." The impatient bus driver agreed with a quick snort.

With great deliberativeness, Heinreich returned Lise's French gesture with two slightly longer, slightly more pregnant kisses of his own, one kiss to each cheek. Lise's hair flowing in the breeze, the warmth of her skin, made Heinreich so want to kiss Lise's lips. But as much as he wanted to—and he almost did Lise noticed—he stopped short. It was not the right moment to do this, and by stopping short of kissing her lips, Heinreich decided that it would give Lise something of her own to pine over. Lise blushed. She understood and smiled. The two of them, who had had crushes on each other for nearly four years, who had suspected as much about the other, and had never felt it proper to openly reveal their feelings, had now finally and precipitously declared their feelings before the world, without a single word having passed between them. "I will miss you," Heinreich then said to Lise as he climbed aboard the bus. "And I will miss you," Lise had responded in her turn. Now that Lise, the one who had looked more beautiful than ever that wonderful day, was lying in bed clinging to her life hundreds of damned, clanking kilometers away.

***

When Heinreich finally arrived at the Berlin train station, he hailed a cab in the rain and proceeded directly to the Reber household. Stepping into Lise's room, he was shocked to see how emaciated she had become. Lying in her bed in a deep sleep, Lise's effervescent constitution was gone. The girl was pallid, her skin tone had a tinge of gray and her eyes were sunken in. She had become a sallow, rail-thin fragment of her former self, and, according to the haggard looking Max, all this, incredulously, had happened in a matter of days.

To his credit, Heinreich kept his cool for Lise's sake. The next day when she had regained her lucidity long enough to realize that he was sitting in a chair by her side—he had sat there by her side half-asleep, half-awake throughout the stormy night—she smiled at her vision not sure whether to believe her eyes. Heinreich, for his part, took hold of her cold, little hands to assure her that he was no phantom. "Mein dear, dear Lise. I go off to France and look what becomes of you. Now that I've returned, you shall soon be convalescent, and I shall take you flying to a cold, cold glacier no matter what mein mother or your doting parents might say. How do you fancy that?" Lise shut her eyes. It was Heinreich beside her. "I'll try... not to keep you grounded... too long," she half-mouthed out between labored breaths before her eyes faded away.

***

It was one of Max's colleagues, Dr. Spalding, a balding, stoop-shouldered physician who specialized in obstetrics and gynecology, that finally figured out what it was that had attacked Lise. By studying cultures grown from her urine, then samples taken from her vaginal area, he determined that a bacteria, staphylococcus aureus to be specific, had invaded Lise's body and released toxins that lead to necrosis of her ovaries—that's where she had hurt the most. The infection, which had started with her flu, had then somehow, he conjectured, worked in collusion with her body's first period to infect Lise's ovaries. Dr. Spalding was correct. It was toxic shock syndrome, as it would become labeled in 1978 in children and in 1980 in women, that had ailed Lise, and it had entered her body through an extremely rare port of entry compared to the famous cases associated with tampon use in which many women died.

Though Sir Alexander Fleming had discovered penicillin in 1929, the actual use of penicillin to treat bacterial infections would have wait until the 1940's, far too late for Lise. In the middle of 1930, Max and Dr. Spalding had neither the promised bacteriophage antitoxins being researched since 1911, nor modern antibiotics with which to treat Lise. To save her life, they both concluded, Lise would have to have surgery as soon as possible to remove her ovaries and whatever else was infected inside of her, and the sooner the better, before the bacterial colony and its toxin could overwhelm the rest of her frail body.

In a whirlwind of activity, Lise, febrile, shivering, constantly slipping into unconsciousness, was ushered into the private hospital near the center of Berlin where Max and several other of his premier colleagues practiced. Heinreich was allowed to join them, but he was forced to stay in Max's spacious, third floor office. There, feeling completely useless, staring out of the window in a hollow daze into the rain-soaked crowds below, the young man would have to keep the good doctor's various skeletal displays company until given further notice.

Within the hour of her arrival, Lise was prepped and put under the knife. The tall, muscular, blond-haired surgeon, who was a relatively new addition to the hospital's ranks, did not quite understand Dr. Spalding's theory of a bacterial colony releasing some kind of necrosis causing toxin into Lise's ovaries. That, plus Lise's weakened state, made him reluctant to cut into Lise's body. But as Max, the girl's father himself explained it to him, time was of the essence as Lise was clearly dying. "The bacterial colony will continue to reproduce inside Lise's body until it kills her, or until it is removed from her body. Please man! Do what we ask of you."

Lise's surgery was touch and go. The infection had invaded not only one of Lise's ovaries as Dr. Spalding had correctly surmised, but her uterus and oviducts as well. It was with great lament that he ordered the surgeon to incise all these organs. Lise, if she survived the surgery, would be barren.

After Lise's reproductive organs were put in a formaldehyde filled jar for future study, both Max and Dr. Spalding concluded to their relief that the bacterial infection had been trapped along the peritoneal membranes, thus preventing the bacteria from penetrating the organs of Lise's abdominal cavity. They cleansed out the area as best as they could and had the surgeon close Lise's incision.

Seven tense hours passed before Lise stirred in her private recovery room. All the while she slept, her fever dropped until it reached only one degree above normal. Heinreich, meanwhile, laying on Max's couch, managed to get some much needed sleep during this time. He didn't dream.

When Lise regained the first vestiges of consciousness, Lise's first coherent words were, with eyes that shone through her tears, "It hurts daddy. It really hurts." Max, who was feeling relieved his daughter's fever was breaking, injected her with a calculated dose of morphine—he would do so for the next 24 hours before switching her to other less effective, but far less addictive pain killers.

When Heinreich walked into Lise's private room a few minutes afterwards, his eyes welled up with tears of relief and agony over her pain. Max and Belinda, who had been sitting beside their daughter, sat up and walked to the back of her room to give Lise and Heinreich a little privacy. Heinreich brushed Lise's brown locks back—he had never felt comfortable doing something like this before.

"Thank you dear Lise, for having taught me the French *bise*," Heinreich began after taking Lise's hand into his own. "It is a skill which, I reassure you, I have not put to use except with a very old and very ugly English princess—a childhood friend of mein mother—who sports a huge, hairy wart on the tip of her nose and three crooked teeth." Heinreich's colorful description, acted out for emphasis with puckered lips, made Lise, holding her lower abdomen with her free hand, break into a soft laughter when his lips pressed against her own lips. It was the best first kiss she could ever have imagined.

***

From the moment Heinreich and Lise bid themselves adieu at the Castle Academy's bus depot, both of them knew that things between them would soon change. Lise's frightful brush with death during the closing weeks of the break of 1930 certainly precipitated and catalyzed that change. While Heinreich stayed in Lise's warm, bustling home to help her convalesce, a new and much more intimate closeness percolated between the pair.

By the end of the first week, to her father's joy, Lise managed to take her first delicate steps within the confines of her ample room. Heinreich, standing ready to catch her, held her hands while she struggled to plant her unsteady feet on the ground. The three paces she managed to take, with considerable help from Heinreich, were a painful ordeal that left her drained and gasping for air at the foot of her bed. "How am I to make it back to my bed?" Lise wondered aloud with a pained expression. "Slowly, and with my help," Heinreich reassured her.

A few days afterwards, with significantly reduced swelling around her incisions, Lise, to her great delight, made considerable progress when she managed to leave her bed unassisted and inch her way to the open window at the back end of her room without a helping hand—at the time, Heinreich was at the rail station collecting the baggage he had left behind in his haste to reach Lise. Seeing the sunlight pouring upon the grounds, and feeling the caressing touch of the warm breeze brushing past her face and coursing through her unbound hair lifted her spirits. It had been nearly two weeks since Lise had set her eyes on anything out of doors. Beginning to feel a little queasy, she sat upon the window ledge and let the vista of Olga playing in a flower bed down below soak in. She enjoyed the dulcet scent of her mother's lilacs. When Max learned of Lise's little sojourn to the end of her room, he ordered Lise to try to walk the length of the upper floor on her next constitutional. "The harder you exert yourself," he explained to his daughter, "the sooner you will heal."

Heinreich promised Dr. Reber that he would personally see to it that Lise complied with his request.

The next day at breakfast, with the servants bustling about the long breakfast table, and Nanerl crying, and two year old Olga laughing hysterically because the dog ate the eggs she had dropped, an overwhelmed Belinda tried in vain to tend to both of her babies while Max hid behind his morning paper. Unnoticed in a corner, Heinreich and Lise sat side by side enjoying their biscuits and cream. Half an hour later Heinreich escorted Lise outside her home for the first time since her surgery. He led her into the shade of an old oak tree where a wooden bench lay beside a flowing fountain. He turned to face her and kissed her without regard to the gardener who was busy trimming a hedge.

The remaining days flew by with Heinreich and Lise talking about anything in particular and everything in general on Lise's lengthening constitutionals beneath the warming rays of the summer sun. They did so with haste as time was running out for them. Heinreich would soon be returning to the Castle Academy. Lise, who was still very weak, wouldn't be joining him until her father was satisfied that she was up to it.

When the final weekend arrived, Dr. Reber treated the family to a surprise. He arranged to have a movie played at home. It was the Marx brothers 1929 release of The Cocoanuts—Max loved American made films, especially the comedies—and the movie brought down the Reber household. Even Heinreich, whose English was the weakest within the Reber household, appreciated the comedy and vowed to pay more attention to his English lessons. After the movie ended, Heinreich retired to his room. With Lise sitting on his bed, he packed his baggage. He would be leaving first thing the following morning.

When the cab pulled up to take Heinreich to the bus depot at the appointed hour, Heinreich bid farewell to Belinda and her two baby girls. "Thank you so much for your wonderful hospitality Frau Reber," he said to Belinda while she dried her hands of spilt milk. After Belinda put her small towel aside, she threw her arms around Heinreich and squeezed him hard. She released him a few moments later to wipe away the tears that had streamed down her cheeks. "Thank you dear boy for bringing Lise's sprits up, and make sure to thank your mother for all the wonderful flowers and gifts she has sent us."

Heinreich assented, "I will. I promise." Max, who was off

at the hospital, had said his goodbyes to Heinreich the night before.

Holding his free hand, Lise accompanied Heinreich to the front entrance where the cab had parked. "I'm going to miss you mein sweetheart," he told her.

While the driver walked up to the covered entry-way and collected Heinreich's baggage, Lise embraced Heinreich, kissed him on his cheek and whispered into his ear, "I'm going to miss you too my Liebchen."

Heinreich, kissing Lise's cheek anew, replied, "I will count down the days until I see your happy face walking the corridors of the Academy—and yes, I shall say hello to Hans for you. Please don't take ill again."

Lise laughed as she let Heinreich's hand slide out of her own hand. "I promise I won't."

As the cab drove away, Heinreich took the time to enjoy the passing scenery of Lise's affluent neighborhood. Everything was so clean, fresh and orderly. He felt happy. He had gotten to know Lise much better during his brief stay with her, during their constitutionals and long conversations, than he had in four years at the Academy. I wonder, he thought to himself as the cab entered the busy streets of Berlin, what Hans will think of the change.

***

To her dismay, when Lise returned to the Castle Academy for the fall session of 1930, she noted that things had changed in many disconcerting ways. For one thing, many of the faces that she was familiar with were nowhere to be seen. Everyone in Germany understood that the failing economy was casting dark shadows on Germany's mood. More so than this, the failing economy was having real consequences. About ten percent of the Academy's student body had been forced to quit—to help at home—due to lack of money. Their palpable absence brought the point home that what was being written about the German economy in the daily papers, that people were losing their jobs and going hungry, was real.

Then there was the new militarism evident everywhere, the most obvious sign of it at the Castle Academy being the required use of Nazi uniforms. Heinreich had written Lise about the new black tunics. Given Heinreich's penchant for jokes, she had dismissed the letter until, in his next post, she received a photograph of the Academy's student body standing in formation in their new uniforms by a new bronze statue of Hitler. "Dear Lise, we are all called cadets now," the postcard began. "I look forward to seeing you soon, Cadet Major Reber." The post card was signed Cadet Oberst (Colonel) von Onsager, this being the cadet rank he had been assigned. Only a few select junior and senior upper-classmen could hold the field rank of cadet major or higher.

During its first four years the Castle Academy had taken a far more academic bent than militaristic. It had not, for instance, been much more militaristic than the American Boy Scout organization. Other than eugenics, some occasional Teutonic propaganda, and the mandatory marching to meal formations, no constant dribble of Nazi propaganda had been imposed on the student body. Frieda, despite her husband's wishes and frustrated desires, had much to do with this, especially when the Nazi party was struggling for its survival and she was footing the Academy's sizeable bills.

The attendant rise of the Nazi party during Germany's collapse, which gave the party increasingly more power and resources to foment strife and discord among the classes, and to impose their stilted values on the German peoples, began to make its doctrines increasingly felt at the Castle Academy. To overcome Frieda's interference, the Nazi party had Wolfgang fire many of the old faculty and replace them with party surrogates. The party picked up the Academy's payroll as well. That's when, against Frieda's wishes, the Nazis instituted the mandatory uniform requirement.

Also against Frieda's wishes, under the supervision of a specially appointed Nazi political officer, the Academy's student body was forced to form a branch of the *Reichsjugendfuehrung*, the Reich Youth Leadership to promote and enforce the most stupid elements—as she thought—of Hitler's supremacist Aryan doctrines. Under the Nazis, for instance, flirtation and public displays of affection were suddenly highly discouraged. From the fall session of 1930 and onwards, Academy cadets would no longer be allowed to even hold hands in public.

Wolfgang, at last, began to put together the kind of Academy he had always wanted to build, the one that he and Hitler had dreamed about years back as opposed to the type of place that Frieda preferred. To hell with the Greek classics he thought. It's time for the Germans. Thus it happened that within the few weeks Lise had remained at her home convalescing, the Academy, through its new faculty, the *Reichsjugendfuehrung* and its political officer, changed into, essentially, a kind of Nazi propaganda camp promoting a dismissive attitude to all but Hitler's views.

"Well Lise, what do you think of the new place?" Heinreich asked as Lise settled into her new quarters—as upper-classmen of field rank they would not be sharing their dormitory rooms.

"I think the changes are simply awful," she replied. "And I despise the uniforms. They are so…so morbidly black."

Hans, who had helped carry Lise's things, disagreed. "I

don't think so. I think the place was too lax to begin with. Now that the deadwood is gone, we can proceed apace." Both Heinreich and Lise, exchanging a hidden glance, kept their peace. "I will leave you two alone as I have much to attend to," Hans continued. "Go to bed early Lise. We rise much earlier than we used to and have many more political ideology classes to attend to."

After Hans left Lise's dormitory, Lise asked Heinreich what Hans' remarks and attitude were about. "I honestly do not know," Heinreich replied. "Hans came back angry, moody, and he has been swallowing up this new Nazi silliness. This means, unfortunately, that you and I will have to keep the developments between ourselves secret, or, I'm afraid, Hans may cause us problems."

Lise, even more disheartened, couldn't believe what she was hearing. "What about your mother in all of this Heinreich? I can't believe she is behind these changes."

Heinreich shrugged his shoulders and raised his eyebrows. "Of course not, but she no longer has the influence she enjoyed." Heinreich explained to Lise the new situation that his mother found herself in, and how his stepfather was finally getting to run the kind of Academy he had wanted to run from the outset

"This is awful," Lise concluded with a sigh. "I had thought Director Boltzmann was in line with your mother's feelings. At least you and I have each other during our last year here. As for Hans, I'm certain he will come around to his old self."

Heinreich replied that he did not think so. "I fear that Hans may be gone for good Lise. When the call for *Reichsjugendfuehrung* volunteers came around, Hans was first to put his name on the list. He was made a squad leader and has been reveling in his new black uniform with that silly

Reichsjugendfuehrung cap. The new students...er cadets I mean, fear him, and do so with good reason. Hans is quite generous with issuing demerits and the new, harsher punishments that have been authorized."

Lise, tried to imagine Hans in this new light, but couldn't accept this horrid turn of affairs. "I wonder if Hans' father, who has always been such a staunch Nazi, has anything to do with this change in our dear Hans. I should talk to him about it." Lise's uncanny intuition hit close to home. Hans was indeed reacting to his father's Nazism. For beating Constance, and for calling him a rat and Jew boy, Hans was going to be a better Nazi than his stupid father ever would be.

"At least," Heinreich concluded, "this is our graduation year, and we know many quiet places where a boy and a girl can enjoy a quiet conversation."

Lise laughed as Heinreich, after closing the door, approached her. "I've been thinking about kissing you for so long now it hurts," he told her.

***

It was several weeks later that Lise made the greatest mistake of her life. Hans received a letter from his mother Constance that she could no longer bear to stay with Himmer, and that she was returning back to her parents home. *"Don't worry about staying at the Academy,"* she wrote. *"As long as I stay married to your father, he won't dare cause you any trouble. I told him that I would make a divorce an absolutely hellish affair for him. Thanks to your grandmother's quick thinking, I have photographs of him asleep in a drunken state among other kinds of photographs which I would personally show to the local party leaders, and even to his own father. The fear that such scandal will cost him dearly has persuaded him that I would best off staying with your grandparents. Be good to yourself and try to be happy. You are nearing graduation, and soon you will be a man, a very good and loving man, even if a bit young in years. Remember dear Hans that I love you."*

Hans didn't have to let his imagination run wild to read between the lines. Constance had obviously been beaten again, possibly more than once. All he had to do to imagine her bruises was to close his eyes and recall the first time he had seen his mother's swollen face. Hans wadded up his fists and struck a brick wall several times bloodying his knuckles. The frustration and anger he felt, which he could barely contain, drove him to seek Lise. The nearly fourteen year old boy needed someone to talk to, and he didn't feel like expressing a weakness to any other boy, especially not to Heinreich. Hans knew full well that he could trust Lise implicitly, and he had no doubts that she cared about his well being. When he caught up to her in a basement hallway by a chemistry laboratory, he asked her to meet him at an isolated spot near the new parade grounds. He told her to come alone after the dinner formation.

Lise, by the wild look in Hans' angry eyes, could tell something was terribly amiss. "Are you well Hans?" she asked him.

"Yes...yes," he replied, "just please come alone and I promise that I will tell you all."

It rained that afternoon and there was a chill in the blowing mountain breeze that washed away the sun's heat. Hans looked smallish and cold beside Lise as they walked to Hans' appointed location. Lise sat herself at the base of a tall pine and remained silent while Hans made certain no one was around. When he was certain of this, he began to spill his guts about his and his mother's life with Himmer. His angry, anguished words shocked Lise. Unable to restrain himself, Hans began to sob after a point, the more strongly so when Lise, ignoring the rules of no public displays of affection, cupped his hands into her own. Hans, overcome with grief, did not object. "I worry so about my mother," Hans concluded with a puffy, red face.

Lise didn't know where to begin. "I am so sorry to hear about what you and your poor mother have suffered through for so long," Lise told the teary-eyed boy after a pause. "But dear Hans, your mother is okay for now isn't she?"

Regaining a bit of his composure, Hans again looked around to make sure nobody had come near. He answered Lise's question with a touch of sarcasm. "I'm sure she's not going to die, but I'm sure she's not at all doing well either."

Lise, remembering some of her mother's stories of past trials and tribulations, tried to raise Hans' spirits. "Hans, even though you are terribly anguished now, and things are going badly, things do turn around. I promise you that. Your mother is doing the right thing, and after you graduate, I am certain she will divorce your father. You must encourage her to do this so that she may begin anew."

Hans' angry expression turned to doubt. "How would you know about the future Lise? Just because your own family is so perfect?"

Lise squeezed Hans' hands. "I have not known strife like you clearly have my dear boy, but my own mother's parents once had to leave Bonn back in 1879 because of so much stupid...." Lise, recalling her mother Belinda's constant admonitions, and Germany's current state of inflamed anti-Semitism, stopped in her tracks.

"Because of so much what Lise?" Hans broke in after losing patience.

"Because they were innocently caught in a corrupt

business deal," Lise replied inventing a lie. "It was shameful, truly it was the way my grandparents were cashiered from their community," she continued, "but then my mother's parents, through hard and *honest* work, were able to restore their good name in Berlin. It wasn't easy mind you. But they persevered. Based on what you've told me about your brave mother, and what I know of you, I have no doubts for your or her future."

Hans accepted these kind words with a little more grace.

When Lise was alone back in her dormitory, she made a mental note to be more careful about revealing her family's past. She hated the ever present, ever growing anti-Semitism that was pitting German neighbors against German neighbors. Like her father, she believed the whole thing would blow over as had already happened more than once since the times of her grandparents youth. There were times when she wished she had never stumbled into a box containing all manner of Jewish objects, an old Torah, several texts written in Hebrew, a Star of David, and so forth in an old, dusty box put away in the attic. She was seven at the time of the incident, and had been looking for her old baby clothing to dress her dolls. Events close to home, however, made it impossible to stick her head in the sand. At her own father's medical practice, the very same young surgeon who had excised the infection out of her belly, had been let go when none of the clinic's German patients wanted anything to do with him. His picture had come out in a prominent newspaper marching at the head of an anti-Nazi rally, and with that his career at her father's hospital was ended despite Max' protests to the hospital administrators that it would all blow over soon enough.

When Lise recounted Hans' troubles to Heinreich later that evening, he was flabbergasted. Before turning in for the night, he penned a quick letter to his mother. He was sure that Frieda—keeping herself in the background, a skill she was good at—would see to it that nothing could jeopardize Hans' career at the Castle Academy." It gladdened Lise to proof Heinreich's letter the following morning. "You're a good friend dear Heinreich. You write about Hans with so much feeling."

Lise's words made Heinreich feel uncomfortably self-conscious. "I'm sure Hans will come out of this, but we have to be there for him. At least now I'm beginning to understand why he has come back so changed, so angry, and so obsessed with this new Nazi craze. I'm certain I wouldn't feel and do much differently if I were him."

***

As the weeks passed, things did not get better with Hans. Influenced by the teachings of the Academy's new, rail-thin, chain-smoking political officer, he grew angrier and nastier by the day. The political officer, Franz Stern, in his turn noted this resonance with the boy. *"Hans,"* the avid fanatic of Hitler wrote in a report, *"should be recommended for membership into the SS. Last week, I witnessed the boy tear into two first year cadets, making them weep in their boots. This week one of them quit, while the other seems to be making the appropriate adjustments. Hans is providing the kind of cadet leadership that is helping Germany weed out the chaff. Keep an eye on this one. He will go far, especially if groomed for duty in the SS."*

History would record the SS, or the Schutzstaffel, as Hitler's most feared and reviled organization both within and without Germany. The "new" Academy, for its part, would grow some of the Schutzstaffel's most dastardly members, male and female alike. Frieda, hearing the reports from her husband about the "wonderful" changes in Hans' personality, was horrified, but, no longer having the influence she once had, she was powerless rid the Academy of Herr Stern. Frustrated, she left for Paris. Before leaving, Frieda wrote her son a note. *Dear Heinreich, I'm certain you and darling Lise are as appalled at the turn of events at the Academy as I am. Mercifully, as senior ranking students—I shan't mention the other term—you will be able take your leave from the place most weekends. The keys to the house are, as always, available. Given its proximity to the Academy, and Wolfgang's recent preference to stay put at his beloved Academy, I suggest that you and your darling avail yourselves of it. Ciao.* "Your poor, sweet mother is a dear," Lise told Heinreich after reading the note.

***

Things grew more intense between Lise and Heinreich as their last school year progressed. When Heinreich learned he had been accepted to *Kriegsschule* Potsdam, a prestigious military academy for officers, and Lise learned that she had been accepted to the equally prestigious Technische Univeristät Berlin, both of them were thrilled for each other. For starters, Potsdam was quite near Berlin, only an hour away by express rail. Hans, by virtue of his precocious age and academic excellence—not to mention his growing connections within the SS—had an open admission to anywhere he might have wanted to go.

The much anticipated acceptance letters for Lise and Heinreich had come on a breezy Thursday in early October. Regardless of how close Potsdam was to Berlin, it very quickly dawned on the young couple that they, except during breaks and short vacations, would soon be living their lives apart for long years to come. Moreover, while being a first year cadet at the Potsdam academy, Heinreich, reading the booklet he received for new cadets, read that apart from mail, first year cadets would be rendered incommunicado from the rest of the world for the period of a year. The only exception would be a short break for Christmas. "According to the booklet," Heinreich read to Lise feeling a pit growing in his belly, "it is supposed to make me into a man."

Lise snatched the booklet from his hands to confirm the ridiculous policy for herself. Sadly, it was not another of Heinreich's jokes. "I suppose," Lise said returning the booklet with sad resignation, "that we always knew that before we could start our lives together we would first have to go our separate ways. I just never believed it." Needless to say, once alone within their respective dormitories, their new and ineluctable reality made for a mix of excited and troubled dreaming.

By Friday, with neither Heinreich nor Lise at ease in the increasingly oppressive airs blowing at the Academy—they were outraged at not being allowed to even hold hands in public—they decided to leave for the weekend. It would be their first ever weekend alone. Naturally, the first place they headed to was the nearby von Onsager summerhouse sitting in the foothills of the Alps. Fortunately, the fortress, as Hitler liked to call it, was only thirty minutes away by cab. "We are lucky to have somewhere to go," Lise declared as she stepped into the cab.

"Rank has its privileges," Heinreich replied without feeling the least bit of guilt as he took a seat beside her.

"Your parent's house is kind of quiet without your mother," Lise remarked when they stepped into the empty house.

"I have little doubt that she is doing fine in Paris," Heinreich reassured Lise. "Besides, I kind of like having the house to ourselves. Don't you?"

Lise, putting her things down, assented with a cheerful

yes. "And I won't say," she continued as she began opening the many closed curtains, "that I lament the absence of your stepfather either. Pardon my saying this, but Director Boltzmann can keep his Academy to himself."

"Just listen to the sound of the wind running through all those pines," Heinreich interrupted as he walked over to Lise. "Isn't it so much better than listening to all those damned Nazi chants we have to sing as we, and everyone else for that matter, march about?"

"It feels refreshing. I can hear myself think," Lise replied while walking over to a wall covered in family photographs.

Heinreich, looking on, noted that Lise's black uniform, especially the skirt, didn't despoil her beauty one iota. "You know Lise, you are about mein mother's size. You should look in her room and pick something out. She has quiet a lot of stuff in that closet of hers. I'm sure you won't have any difficulty finding something suitable. Get something warm. The breeze is kind of cool already. We can go out for a brisk walk. I'll pack us a lunch. When we return, I will start us a fire. Let me call Berta and ask her to prepare us a dinner by, say, 6 o'clock this evening. I'll tell her to leave it in the oven. She'll be happy to be cooking again."

Lise put down a photo of Heinreich as an infant she had been looking at. "It sounds like a wonderful plan my dear sweetheart."

***

To Heinreich's delight, the morning's winds had abated outside. Patches of fog hanging low about the trees were burning off in the rays of the rising sun. As the wispy layers of vapor rose from the ground before fading away into the atmosphere, they reminded Lise of living specters, adding an air of mystery and enchantment to the humid forest. She clasped Heinreich's hand as they started for one of Heinreich's favorite childhood places, a quite, calm lake located about ten kilometers away along a trailhead starting from his house. Heinreich planned on taking his sweetheart for a day of rowing.

"Did you ever think you would actually grow up one day?" Lise asked Heinreich as she stepped over an exposed root.

"Not much," Heinreich confessed. "I've always supposed the day would come, but I haven't thought about it much, until now that is, with you and I going our separate ways. Clearly, I'm excited about going off to become an officer in the army reserves and a pilot for Lufthansa." (Unlike Heinreich, who, thanks to his parent's deep connections, was aware of the real nature of Lufthansa pilot training, very few Germans were aware that their country was building a secret air force, in part by training its pilots as airline pilots. Until it was too late, neither the English nor the French made the connection either.) "But I'm saddened, and even a bit afraid at the thought of not having you nearby."

As a squirrel crossed the couple's path, Lise adsorbed Heinreich's words in a moment of silence before expressing some of her own deeper thoughts. "When I thought I was going to die, I was saddened of course, but I was also afraid… afraid not of dying or of death…but afraid for the pain my parents would feel, and for my not being around to help them in their old age, and not being around to help my baby sisters, and, most of all, not being around to be with you."

Heinreich stopped Lise and put his fingertips to her lips. "I never thought you were going to die," Heinreich told her. "Not for one minute did I believe it." He was lying.

"The world is changing so fast," Lise continued along a new line of discourse. "The world is being electrified, covered over in a web of telephone and telegraph lines, shrunken by ever faster ships and trains and automobiles—"

"—And don't forget planes," Heinreich interrupted.

"Yes, and planes too," Lise added. "I feel as if humanity is arriving at its Golden Age. I suspect machines will soon be doing most of our work, freeing us to think, or they may even think for us, or we may one day join ever more capable machines to our own bodies. We may just go on changing and changing at an ever faster pace."

Heinreich chuckled at Lise's intensity. "Do you think people, like Hans insists, will go to moon?" he asked her.

"Dear Heinreich, no one living in this forest land a hundred years ago would have asked that question, except perhaps a lone loon, but now, as Hans points out, it is plausible given the rapid development of modern rocket science. Wouldn't you want to go to the moon?"

This time Heinreich burst out laughing. "What has gotten into you mein dear? Surely I would like to go to the moon and look back at the little ants crawling all over the Earth. But I just want to be happy, fly, make airplanes for a living—like I've told you so many times—and get to do the things mein father never did. Someday I want a complete family, a wife, and adopted children if needs be. I want to see the babies grow up not wondering what their dead father might think of them."

Hearing these words saddened Lise. She would never be able to have her own children. She changed the subject. "As pretty as this forest is, I can't wait to get to the lake."

When Heinreich and Lise reached the lake, they were pleased to see that only a few other people, a family here, a couple there, were out and about enjoying the morning scenery by the placid waters. "Mein father—mein mother tells me— used to love it here when he was a boy actually living in his father's castle, our very own Castle Academy. They would have to drag him back home."

After a few minutes of surveying the scenery, the young couple made their way to the von Onsager's boat shack toting an old bottle of wine, a couple of glasses, salted crackers and a portion of aged goat's cheese.

Lise helped Heinreich unfasten the small boat's moorings inside the weathered shack. The boat, which had belonged to Karl von Onsager as boy, was an old rowing boat with enough room to fit four adults, if snuggly. It had recently been given a new coat of dark green paint and sported a flat, square umbrella sitting closed towards the backside. "I can imagine your father loving this place. Sorry for me to say this Heinreich—and it was very generous of your mother to donate it—but that castle, big as it is, can be rather dark and confining."

Heinreich agreed. "I'm glad I'm not the only one who feels that way. Shall we go mein lady?" Heinreich clambered aboard after Lise was safely sat down. After a good push-off against one of the mooring posts, the young pair were off into the waters beneath the early October sun.

While Heinreich rowed the old boat away from shore, Lise nestled herself into Heinreich's body placing her head upon his shoulder. It was cozy aboard the small boat, but the couple fit nicely together. Of the few other boats which were out and about the lake, none of them were near enough to bother Heinreich's and Lise's peace.

For about an hour or so, neither Lise nor Heinreich spoke much as they ventured out further from shore. They absorbed the pretty sights of the mountain lake and soaked in each other's warmth. When the sun got too bright, Heinreich opened the square umbrella at the rear of the boat. While he did so, Lise opened their bottle of wine, and afterwards spread the goat cheese on several crackers. It was a little past the noon hour, and the two of them were hungry after their ten kilometer hike up to the lake—literally up 350 meters in elevation from where they had started, mostly along a path between two sheer walls of bare-faced rock.

"Are you ready to lunch?" Lise asked as Heinreich poured her a glass of merlot.

"I could use a little something," he replied smelling the odor of goat cheese.

"That's a good merlot," Lise declared a moment later. "Your father certainly knows how to pick his wines."

After lunching, Heinreich and Lise began talking up a storm—mostly about fond childhood memories—as they worked their way through their bottle of wine. The more wine they consumed, the less they talked and the more they kissed. Distracted in each other's voices and kisses beneath their umbrella, they didn't notice the heavy rain clouds rolling in until after Heinreich, breaking from Lise's lips for a breath, noticed them

"Oh mein God!" he declared. "We had better be going Lise. Look at the storm that is fast approaching!"

Lise couldn't believe the size of the clouds that had built up behind them—she had forgotten that weather could build up fast around the Alps.

"We have a good three or four kilometers to row to shore Lise. Look about. No one else is out here except for us stupid boneheads."

***

Sure enough, the first drops of rain began falling while Heinreich and Lise were still far from shore. As the winds began to whip, Heinreich was forced to close the umbrella, which was acting like a sail and pushing the boat in the wrong direction. It was cold rain that impinged on their skins, and it didn't take long before Lise and Heinreich began shivering, but this only encouraged them to row more vigorously. When they finally reached the shelter of the boat shack, the rain stopped as if it had been shut off with a valve. Lise and Heinreich, now chilled to the bone, found it terribly amusing.

Fortunately for the couple, the von Onsager's kept a stash of weathered, raggedy clothes in the old shack for such occasions. When Lise saw Heinreich in his tattered clothes, she burst out laughing yet again. "Why you look like an American hobo." Heinreich bowed. "Why thank you mein pretty, and I should say, you look ravishing in those baggy pants and straw hat." Lise curtsied.

After Heinreich and Lise collected the remains of their lunchtime victuals, they made their way back down the trail to the summerhouse. Despite darkening skies which threatened more chilling precipitation—snow perhaps at their elevation—the two of them were in cheerful moods. "Looking at mein watch," Heinreich broke in, "we should make it home just in time for dinner."

Lise noticed she was hungry again after hearing the word dinner. "That would be wonderful," Lise replied. "I am famished and cold."

About a kilometer shy of the house, a cold rain poured on them again and they began running down the slope hand in hand. By the time Heinreich and Lise got indoors, they were sopping wet and chilled to the bone for the second time. "I'll start the fire," Heinreich said, while Lise went off to retrieve some dry clothing for them. "I'll see to our dinner shortly," she called out from the hallway.

After Heinreich started a fire in the house's main fireplace, a massive work of stone, he suggested to Lise that they dine by its warmth. Lise in turn suggested they shut off the electrical lights and instead enjoy the light of the dancing, crackling flames. The large window beside the enormous fireplace also afforded the couple a wonderful view of the mountains disappearing into the imposing storm clouds far into the background. Berta, the family cook who lived nearby, had prepared the "children" a hearty meal of *Spanferkel* (suckling pig) with a side dish of sauerkraut before retiring to her own home and family. "My!" Heinreich remarked when he opened the various trays, "our good Berta outdid herself again."

With the lights turned off and the food spread out on plates and trays before Heinreich's fire, the young couple sat themselves down to dinner and, forgoing their forks, picked hungrily at the *Spanferkel* with their fingertips. Outside the storm opened up with a spectacular burst of fireworks, tossing lightning bolts and raging gusts of wind in every which direction. At one point Lise began to feed Heinreich morsels of the suckling pig, and Heinreich reciprocated until it was all gone.

"That was positively wonderful," Lise declared. "You will have to convey Berta my thanks."

Heinreich patting his filled belly agreed. "Of course," he replied. "My compliments to Berta."

A lightning bolt struck nearby, startling Lise with its sudden thunderclap. The mood between the young couple grew solemn. After the echoes of the resounding thunderclap died away, Heinreich remarked that he hoped no Alpinists were trapped on the mountain. "One time my stepfather and I got trapped up there, and I thought for certain I was going to die." Heinreich turned to face Lise and took her hands. Her eyes, reflecting the light of the flames dancing in the fireplace, were aglow. "When I thought you were going to die Lise, I thought I was going to die," he confessed.

Lise didn't quite know how to reply. She didn't need to. Heinreich pulled her towards his lips and kissed her softly. Slowly, very slowly, Heinreich and Lise began falling into each other's breathing, and into each other's passions. The progression to more intimacy flowed like a stream following a course of its own ineluctable making. When Heinreich began to kiss Lise's neck, Lise undid the lace of her dress. Flushed with heat, she wanted to feel the touch of his lips and the kiss of his breath run down her body. Cupping a bare breast in one hand, Heinreich's lips were drawn to the nipple of the other. For the first time in his life he grew aroused in the presence of a young lady. Lise, feeling Heinreich's arousal brush against her body, reached below and took it into her hands. The poignant feeling of Heinreich clasped in the palm of her hand made Lise's blood boil. Between passionate kisses, she undid Heinreich's buttons, removed his shirt, and traced a string of kisses down his neck towards his chest. He was a beautiful young man she thought, tall and athletic, very handsome, very caring, and loving, and funny. I love him, she kept thinking to herself as Heinreich finished removing her dress.

When Heinreich finally had Lise resting on her back, a powerful instinct of longing took over her body. When he entered her, it hurt—she had known it would hurt the first time—but his presence within her also felt compellingly good and right. Heinreich, looking down at Lise's face and the whole of her breathing figure running down below him, at once open to him, at once wrapped around him, holding him, swallowing him in, took a moment to close his eyes to store the memory of this moment of magic deeply into the corners of his mind. When Lise wrapped her arms around the small of his back, Heinreich knew what to do, and whilst the storm raged outside the window, a storm of their own raged inside their minds and between their grasping, writhing bodies. When neither one of them could hold out any longer, when every bit of their strength had been used to exhaustion, leaving Lise and Heinreich trembling, he collapsed, folding himself into Lise's open arms. For a long while the lovers remained this way, embraced, entwined, lost in each other.

"I think we've grown up a little bit," Lise finally remarked.

Heinreich kissing the tip of Lise's nose, agreed.

The couple spent the remainder of the stormy night wrapped in a blanket beside the fireplace. Twice before sunrise they fell asleep to the respiring rhythm of their warm, nude bodies. Twice before sunrise, they awoke to make sweet love again.

***

When Lise and Heinreich returned to the Castle Academy early the following Monday—literally at the break of dawn— they were profoundly changed people relative to the ridiculous goings-on of the place.

No one except Hans noted that something had changed between his two friends. He suspected they had finally talked to each other about their feelings, and had reached a quiet understanding to control themselves. He was proud of Lise and Heinreich for not making a public spectacle of themselves. Whatever was going on between them, they were being discreet. That very weekend in which Heinreich and Lise had disappeared off campus, Hans, who also had the privilege of off-campus weekends, had busied himself writing up the infraction of a junior cadet for having patted a girl on her behind after a mixed sex sports match. Such barbaric, animal-like displays were for American niggers or rat Jews, not Aryans. Lise and Heinreich were shocked when Hans proudly related to them what he had done. The two of them detested the person Hans was becoming, even if the boy had good reason to be angry at the world.

As the months rolled by into November and December, Heinreich and Lise found the way Hans abused his cadet authority over his juniors to be increasingly intolerable. Hans, apparently, not taking pity on any of the stragglers, had forgotten the help he himself had received from Heinreich during his first year. With his new Nazi-based perspective, anyone, boy or girl, falling behind was a contemptible rat as he liked to put it, the word rat becoming a favorite of his.

Without mentioning any specific names Heinreich wrote Frieda an alarming letter a few weeks prior to the Christmas break of 1930. *"The Nazi craze at the Academy is making little Napoleonic monsters mother. This whole Aryan business is turning ugly. For my part, I can't wait to get out of this loony bin. This goes for Lise. We look forward to seeing you soon. I would love it if we can see the Rebers in Berlin during the winter break. Perhaps we can dine with them Christmas Eve. If you can read between the lines mother, I don't think I could stand it being apart from Lise on such a special night."*

When Frieda read her son's not so thinly veiled letter in Paris a few days later, it brought a huge smile to her face. Dismissing all the depressing stuff about the Nazis, all she could think was *finally!* She was filled with joy for Heinreich and Lise. One day my son will marry that wonderful girl she told her finely toned Italian companion. The champion swimmer offered her his sincerest congratulations. Frieda's only sour note was that she knew Lise could never give Heinreich a son or daughter of his own.

***

Christmas break went a little beyond what Heinreich had wished for. As had become customary since Lise began attending the Academy, the Rebers combined Lise's birthday with the Christmas holidays. They would normally pull out all the stops, but on this particular break they exceeded themselves, renting a large dance hall and inviting half the town for the joyous occasion. They wanted Lise's eighteenth birthday celebration to be one she would never forget. Frieda, with Belinda's permission, hired the Berlin Philharmonic to perform at the birthday ball. She wanted nothing but the best for Lise. Many a prominent Berlin family—many of them patients of Dr. Reber—delayed their vacations to attend the event not only for the sake of Dr. Reber's daughter, but to catch a glimpse of, or even have a word with the Baroness von Onsager. Lise's birthday party threatened to become the social event of the year, and became the talk of the town during the preceding weeks.

"My what a beautiful young lady you've become," a beaming Max proudly stated as Lise entered the salon in her long, ballroom dress escorted by Heinreich in a black tuxedo. A wave of applause, started by Frieda, filled the dance hall. "Isn't it nice to see our young ones out of those horridly drab uniforms," she whispered into Wolfgang's ear trying to irk her distant husband, but Wolfgang didn't even flinch. Other things were occupying the man's mind that Christmas.

Hitler, feeling confident about his rising political prospects, had sent Wolfgang a letter asking him if he would like to design a new ministry of education. It was a great honor and opportunity to rise in the party ranks. Deep in his guts, Wolfgang was certain Germany was going places with Hitler at the helm. The man and his party had won a resounding victory, taking 6.5 million votes in 1930, and there was no sign, as long as the economy continued to crash, that the momentum of Hitler's movement was abating. Of course, knowing his wife's feelings all too well concerning the new, more aggressive directions the Nazis were taking, Wolfgang kept his considerations to himself. Frieda abhorred the vicious way the Nazis used their acid propaganda to appeal to the anger and fear of the general German populace against the Jews, the Communists and the Social Democrats. Though she herself felt the shameful sting of the November 1918 armistice, and despised many of the elements of the Versailles treaty, she didn't think of those parliamentarians who had been forced to sign the treaty as the condemnable criminals the Nazis made them out to be. Above all, no one, she thought, deserved to be beaten or killed for their political affiliations and convictions, an activity which was becoming commonplace under the Nazis.

Little butterflies in China, unnoticed for what they were, were flapping their wings around the world in 1930. The first known use of optical coincidence cards for searching literature was performed by R. Preddek of Germany that year. In the United States, IBM was preparing to introduce its new electro-mechanical circuit design principle in its multiplying punch device. Also of note in 1930, Paul Dirac predicted the existence of anti-matter, and Walther Bothe discovered neutral rays. These rays were later identified as neutrons, the critical players in nuclear chain reactions and atomic bombs. The reach of mankind's power was rapidly expanding. So too was the universe, as Hubble discovered in 1929. The only rule it seemed, if there was one, was that nothing was standing still, and that change itself was always changing.

At the Fritz household, Himmer, alone, drunk himself to sleep on Christmas Eve. Hans and his mother Constance, on the other hand, enjoyed their time at his maternal grandparents house. The Heisenbergs always knew how to have a good time. They enjoyed the Japanese candies that cousin Werner had sent from Tokyo, and Hans, above all, was relieved to learn that Christmas that his mother had finally decided to leave Himmer for good. Constance promised her son that she would proceed with filing formal divorce papers by the coming June, after his graduation. Now that Hans was close to finishing his work at the Academy—he would be graduating in June as a prodigy of merely fourteen years—Constance began believing that she could continue with her own life. She was, after all, still quite young, and hadn't given up on the notion of having more children with some good man.

***

The 1931 spring session of the Castle Academy—it would be the last session for Lise, Hans and Heinreich—began with excitement in the air. The senior class would be the very first graduating class, and there were great expectations for the continued success of its students at universities and military academies throughout Germany, and even, in some cases, abroad as far away as America. The Nazis were eager to show off the best of their very best, and spread their peculiar ideas wherever sympathetic ears might be reached. So too were many of the graduating cadets who had, by this point, fallen prey to the Nazi's abominable indoctrination.

After Frieda left to tour South America in late January, Heinreich and Lise continued to have their weekend rendezvous at the nearby von Onsager household. Director Boltzmann, ever present at the castle, persuaded Hitler to allow him to wait on accepting the ministerial position until after his beloved Academy's first graduation. Being clever, Wolfgang flattered Hitler by asking the future dictator to give the first ever convocation speech of the Castle Academy's history. Hans, meanwhile, continued to become the ever better Nazi, short and black-haired as he was notwithstanding. Jews, never mind that he didn't ever come across any in his protected life, became his greatest and most despised enemy.

Before anyone could believe it, the long awaited, and even slightly dreaded day of graduation finally arrived. The weeks of practice for the march-in-review were finally to end. All of the Academy's cadets, dressed in their sharp black parade uniforms, excited and nervous, awaited with anticipation the filling of the review stands by hordes of proud parents and scores of Nazi dignitaries. To every cadet's relief, the parade-in-review started without a hitch beneath blue skies when the appointed time, marked by booming cannon fire, arrived. As a military band played marshal scores to the beat of war drums, the cadets proudly marched past the review stands with their chests thrown out. In response, flags with swastikas waved from the stands, and clapping parents snapped photographs while salutes were exchanged and honorary medals were awarded to outstanding cadets—Hans received the Academy's gold SS medal. Then it was Hitler's turn to speak while primitive blank-and-white movie pictures were shot by well-positioned Nazi propagandists.

With the parents now sitting and their uniformed children standing at attention on the review grounds, all eyes turned to Hitler. He began speaking with a soft-spoken demeanor. "You are Germany. Germany is you," he started nodding his head.

"You are the magnificent flower of your parent's generation," he continued with a slightly raised voice. He raised his finger and pointed heavily towards the cadets arrayed before him. "You will restore their honor, the honor that was STOLEN from them at Versailles by cowards and Jews. I am counting on you to do that...to restore Germany's honor." While Hitler paused to collect his thoughts, Director Boltzmann snapped the cadets into saluting the brooding speaker. "Heil! Hitler," he yelled three times while tossing out his left arm. The cadets, a thousand strong, echoed their director's exhortation with stentorian force. "Heil Hitler! Heil Hitler! Heil Hitler!" The power of the moment brought the people of the review stands jumping to their feet, cheering and applauding the little man with burning eyes standing behind the podium.

Hitler, deathly serious, wiped his forehead with a handkerchief. From left to right, while waiting for the crowd in the review stands to settle down, he scanned his eyes over the thousand strong cadets. Heinreich, standing before the cadet regiment he was in command of, actually felt a twinge of pride running up his spine. It was unexpected, but the man, who once again began to rail against those who had and would harm the fatherland, had that eerie effect on peoples. Lise, standing at attention a few rows away, thought Hitler's speech was too angry and too hateful. Hans, on the other hand, with his gold SS medal sparking in the sunshine, swallowed every one of Hitler's words as if they were morsels of ambrosia.

When it was all said and done, and the caps were tossed into the air, the three who had started off as children, Lise, Hans and Heinreich, sought each other out. They were all smiles. "We lads," Lise declared with joy in her eyes, "we've made it!"

Heinreich, giving caution to winds, kissed her in the sea

of happy cadets. "We did," he agreed as he then turned to Hans to pump the boy's hand. "All for one and one for all right Hans?"

Hans, taken aback a little by the overt emotion of the moment, replied with a single word. "Forever." The boy didn't look too pleased.

"Now Hans," Heinreich broke in, "I would like to give you and your father—the bastard did come today I take it?"

Hans nodded yes.

"Excellent," Heinreich continued, "because as I was about to say, I would like to give the two of you a great surprise." Himmer, himself a rising star in the Nazi party from the Munich area, could not have afforded to miss the graduation with Constance standing by his side feigning marital bless. "Mein mother and I, and mein father, who is—"

Hans stopped Heinreich and said, "who is Director Wolfgang Boltzmann," to the astonishment of both Lise and Heinreich.

"You knew?" Lise asked nonplussed.

"Since when?" Heinreich followed.

"I've known for about a year now, ever since our SS political officer Stern revealed it to me. And for about as long, I've also known about the weekends the two of you spend at the house by the lake. My congratulations."

Heinreich, noting a little ire in Hans' voice, was the first to try to fix any problems of offense by omission. "I apologize if Lise and I have slighted you dear Hans. Lise herself did not know about mein father until after her parents sought out mein parents over the glider incident after…after the break of '28. I didn't want anyone to know who mein father was. I wanted to be treated like an equal. I hope you understand this."

Hans—who had in truth felt very slighted—accepted this

much of Heinreich's explanation. "That was most honorable of you baron," he said with an insincere tone of obsequiousness regarding Heinreich's elevated social rank. "I accept your apology," Hans then dryly concluded after clicking his heels.

Lise, not at all liking Hans' affected attitude, placed her hand on the boy's tense shoulder and tried an apology of her own. "Thank you Hans for being so understanding. I do so love your magnificent medal." She followed her apology with a kiss on Hans' cheek. Then, with a loaded tone which Heinreich understood, she added, "If anyone deserves that SS medal, it is you." Lise was not at all pleased at being spied upon by the filthy SS.

"As I was trying to say Hans," Heinreich broke in trying to cool things, "you and your family are invited to a celebration at mein mother's house, and you, as the cadet with the highest academic marks, and winner of the Academy's SS gold medal, are to be the guest of honor. Hitler himself, mein father saw to it, will present you with another, personalized medal."

Only now did Hans smile. "My bastard of a father, as you so aptly put it Heinreich, will have to eat his damned words in front of me and mein mother won't he?"

"That is the general idea," Heinreich added with a wink.

***

The graduation celebration at the von Onsager's hillside summerhouse went well, as all of Frieda's soirees did. After the dinner victuals had been enjoyed—the best foods from the best of Bavarian cuisine—cigars were dispensed to the men while bottles of Portuguese port and French Champaign were opened. After a few rounds of toasting, Hitler quieted the small crowd and announced with great ceremony the personal medal he was about to award Hans.

With Himmer and Constance standing beside Hans, facing Hitler who stood before them, Hans bowed to receive his medal. "This young lad," Hitler began, turning the beaming boy around to face the crowd, "with still so much more to grow, with his brilliant mind, and his zeal for Nazi ideals, has brought pride to me and to his countrymen. It is my honor to present him the first gold academics medal of the new Reich ministry of education, which, with no further delays, our able Director Boltzmann will lead as Minister Boltzmann."

Loud applause burst out, during which Hitler reached out to Himmer, took his hand and pulled him in close to whisper something into his ear. "I would be careful," Hitler whispered to Himmer while the crowd drew in around Hans, "of whom I called a dirty Jew-boy rat."

Himmer turned ashen pale. "Yes...yes of course my Führer," was the only thing Himmer could say after Hitler released his hand.

"Thank you Frau Fritz," Hitler said turning his attention to Constance. "No boy could achieve what Hans has achieved without an extraordinary mother."

Hans' heart nearly pounded itself out of his chest with pride. He had triumphed over the great injustices and affronts of his hateful father, who now stood so ceremoniously undone. He, the Jew-rat boy, had shown the wife beating Himmer a thing or two about who would never amount to anything at all.

***

Hours later, when everyone had departed except Hitler and the Rebers, who had put Olga and Nanerl to bed earlier that evening, coffee, pastries and brandy were brought out to wind down the evening of festivity. Hitler was in a cheerful mood. "Heinreich and Lise are the kind of Aryans that will make the Fatherland the greatest nation of all times," Hitler declared, embarrassing the young graduates. "That academy of yours, Director Boltzmann—Minister Boltzmann rather, is the finest institution in the world of its kind. If only I had been so fortunate as a child to have attended its like, then where would I be now?"

It was Wolfgang's turn to be embarrassed by his old friend. "Thank you my Führer," he replied with a bow.

Declining the pastry tray as it was passed around by a servant, and not taking anything to imbibe, Hitler excused himself. He would be leaving for Berlin first thing in the morning to attend to matters of state.

The von Onsagers and the Rebers continued with their celebration after Hitler was off to bed. The parents spent about an hour embarrassing the "children" with childhood stories. Lise and Heinreich in their turn related their more amusing experiences at the Academy. Neither the Rebers, Frieda or Wolfgang were expecting the great surprise Lise and Heinreich had in store for them.

"Lise and I," Heinreich began after taking hold of Lise's hand, "wish to go to Paris, alone, before we are forced to go our respective ways."

Baroness von Onsager burst out laughing at Heinreich's boldfaced announcement.

"You are joking right my dear boy?" Wolfgang asked.

"Not at all father. As you well know, I will be locked away at Potsdam for a good while. Lise and I want to be alone, spending as much time as we can together before we part."

"But no!" were Belinda's first words.

"Why not mother?" Lise asked.

"Because the two of you are children," Belinda replied.

"We are not children mother. We are the future of Germany."

Dr. Reber cleared his throat. "Lise, you know very well what your mother meant."

"Father, it's not as if I can get pregnant."

Belinda gasped, but before she could say a word, Lise cut her off. "Mother, before you protest too much, I read your university diary years ago. So if you say no, Heinreich and I will get married as soon as possible. We will do so without any of the pomp or circumstance all mothers dream about."

"But Heinreich!" Wolfgang protested, "you do know this would disqualify you from Potsdam."

"Precisely father, it would."

Belinda turned to her daughter with a look of puzzlement. "I just don't understand?"

"Mother, you only have to understand that I love Heinreich, and that we are serious about each other. Very, very serious mother, and now, or after our studies are complete, we shall marry."

Dr. Reber tried to calm things down. "Since the days I saw Heinreich hovering devotedly over my daughter while she teetered between life and death, then stay by her side while she convalesced, I have foreseen such an eventuality. What might have been a crush, a case of simple young love between our Lise and Heinreich, has grown into much more. I trust this to be true, and I trust them."

"Thank you father," Lise said after her father's remarks.

The other adults took a moment to ruminate over Max's words.

"Heinreich, look into my eyes and tell me that you will really marry Lise if we forbid the pair of you to go to Paris alone. As much as I love you as my own son, I simply will not tolerate an idle threat."

"Yes father," was Heinreich's simple response.

"Really Lise?" Belinda followed up with her daughter.

"Yes mother," Lise replied. "We are serious."

Frieda, placing her hands on her lap, turned to her son and said, "Well my son, you are eighteen years old, and therefore a man, even if only a young man, before the eyes of the law. You can of course do as you well please. All I can threaten you with is your father's inheritance, but you know, of course, that I would never deny you it. Karl would roll over in his grave if I did so. The same applies for Lise. Legally, dear, you are no longer under the protection of your parents. The matter is up to you, not up to us. For now, I suggest we all retire and sleep on it—as for you two, I suggest you sleep on it *apart*."

Frieda let her husband know what she really felt when she and Wolfgang retired to their boudoir. "As long as Heinreich doesn't do something stupid to ruin his promising future, I don't care if Heinreich and Lise run off to Paris. I think it's a stupidly, wonderfully romantic idea. Lise, after all, cannot get in trouble. The poor thing will never be able to get in trouble, and we can't deny that she has been very good for Heinreich. As you can easily remember, the boy did not have a sense of direction before the darling came into his life. She, and flying to be sure, has helped him find a good path. Professor Rudin has related to me several times that Heinreich, without Lise would still be piddling around in basic calculus courses rather than nearly finishing the equivalent of an undergraduate degree in maths. Göring has told me he believes Heinreich can one day revolutionize aviation."

Wolfgang couldn't disagree with these assessments. "Our Heinreich is taking after his father isn't he?" Wolfgang concluded. On this point, Frieda couldn't disagree with her husband.

In the Reber bedroom things were more heated. Belinda was still shocked at the thought of losing her baby, not to mention her grappling with the idea that her baby was, apparently, no longer a baby, but a young woman. Max did his best to reassure his wife that her first baby girl would always be her baby. "Let them go," he told her, "and be happy for them, otherwise forbidding Heinreich from Lise will only make things worse. After the summer they will, as they themselves know, have to go their separate ways while Heinreich is trapped at the Potsdam academy. If they truly love each other, then their love should endure this trial. If not, then no harm would have been done by their having gone to Paris for a little fun. I want Lise, and when the time comes, Olga and Nanerl to be wise and experienced girls, like you were. I don't want what happened to that poor women, Hans' mother, to happen to our daughters."

Belinda didn't accept any of Max's logic. She only resigned herself to it. "At least Heinreich is a very good boy," she concluded.

Naturally, after Heinreich and Lise believed the "adults" had fallen asleep, they did what young lovers running against the world did. They snuck out, got into Wolfgang's car and drove to a nearby park to make love.

***

Their days in Paris were a whirlwind of romance for Heinreich and Lise. The evening café dinners along the Champs Elysée, the early morning strolls about the Seine in the shadow of the Eiffel tower, and all of the rest of the enchanting offerings of the "city of lights" treated the young lovers to days of dreamy laughter and nights of young, joyous passion.

The first thing Heinreich did when Lise and he stepped off their sleeper wagon at the main Paris rail station, was to give her a welcoming in the style of the French. "Bienvenu mademoiselle," he told her after kissing her on alternate cheeks. "I believe—someone once told me—they call this odd but tasty greeting the *bise*. Now you must kiss me once upon each cheek."

Lise laughed. "I think I recall something about this peculiar gesture at some point back in my distant past."

It was at the cathedral of Notre Dame that Heinreich did his most romantic thing. "Can you believe that they finished constructing this place in 1250?" Heinreich asked Lise while he led her past the gargoyles at the entrance. "How many people have passed through these doors? Can you imagine it? All the hopes and dreams of all the young couples who came here to be wed?"

Lise didn't quite understand what Heinreich was getting at. "Unfortunately," Lise broke in, "the most famous couple that ever found love here was Quasimodo and his doomed Mercedes."

Heinreich chuckled. "You would say that wouldn't you. I am familiar with this story. It is one of my favorite Victor Hugo stories along with *Les Miserables* and *Toilers of the sea*, but that is not what I'm getting at."

Lise remained confused. "Now dear Lise, if you would, please do follow me to the Rose window. I want you to see how the colors fall upon you. They say that 650 years of aging has made its effects magical."

Heinreich grabbed Lise by the hand and towed her over to the most famous thirteenth century window in the world.

There, without her expecting it, as the light from the sun poured down upon Lise in a rainbow of colors, he dropped to his knee and proposed marriage to her. "I love you so much Lise. I want you to marry me. Will you Lise Reber marry me Heinreich von Onsager?"

Lise was stunned. Throughout the length of her body she felt the same waves of heat and chills wash over her that she had felt when she and Heinreich had first made love. A feeble, "when?" was as all she could manage to say a few breaths later.

"Soon mein dearest, when I become Leutnant von Onsager. Please, take this ring in acceptance of mein proposal. It belonged to mein grandmother."

Lise looked down upon the old ring. It was beautiful and ornate in the old fashion style. For a moment Lise felt as if she dare not touch it, but something inside her compelled her hand towards the ring. "I will marry you Baron Heinreich von Onsager. I will."

When Heinreich heard these words, he was suddenly beside himself with joy. The ring, which he then proceeded to place on Lise's finger, fit perfectly. He had sized her finger once while she had slept in his embrace a few months prior to their graduation.

"I will wear it always," Lise proclaimed while bringing the ring closer to her eyes.

Heinreich laughed. "It looks beautiful on you making, as I had hoped, all those sparkles in the light of the old cathedral's window." Lise, unable to keep her eyes off the ring, didn't say a word.

"Do you like it?" Heinreich asked.

"It is the most beautiful ring in the whole world," she declared to Heinreich's relief.

"Good. It will be yours until mein mother realizes the ring is missing." Lise looked at Heinreich with alarm. "I jest Lise—truly I jest. Mein mother gave it to me for this very purpose the night before we departed Berlin."

When Heinreich and Lise returned to Berlin, they were happier than a pair of clams. Neither Belinda nor the baroness knew quite how to address their adult children. Frieda, perhaps because her grown up child was a son, was the least stultified of the two women. "I can tell by the glow of your face that you enjoyed yourself," she told her son when he returned home. Heinreich kissed and embraced his mother. "We did mother."

"I'm glad that you and Lise love each other. I take it she accepted the ring?" Frieda got her answer from Heinreich's look alone.

Belinda, on the other hand, told her daughter that she wished she would have waited until she was older to become a woman. "I know you are very intelligent my daughter, but you are still so young Lise." On a positive note, Belinda was very pleased with her daughter's engagement ring. "Its stone is so large," she said after inspecting it under a lamp. "I have only one question for you daughter. Have you ever thought that Heinreich may one day want children of his own? Then what?"

Lise put her hand to her belly. "You asked me two questions mother. Heinreich knows that I can never give him a child of his own. We will adopt children."

Belinda eyes watered. "I wish you hadn't gotten so ill. I'm so sorry for you."

Lise put her arm around her teary-eyed mother. "Don't feel bad mother. I am alive, and I've found someone who loves me."

***

On the way to the Potsdam Military Academy, all pilot candidates were required to attend a meeting with Herman Göring. "Gentlemen," he told the candidates after they had assembled in a large auditorium in Berlin. "There are a few things I have to inform you of before you proceed with your training. First of all, you may consider it awkward to be on your way to becoming officers in the Army, and commercial pilots afterwards. As you all know, the Versailles Treaty forbids us from having our own air force. This is our way of circumventing the will of France and England to keep our nation weak. Our air force, the Luftwaffe, meanwhile is being assembled and trained in Italy and the Soviet Union in secret air bases. The Luftwaffe will not become official until we deem ourselves strong enough to defend ourselves. This is why you will be commercial pilots with Army commissions flying in the 'black air force.' The good news is that we believe your officer class will be the first to enter the official Luftwaffe and train in bases here at home." The existence of the German air force, or the Luftwaffe, would not be officially declared extant to the world until March of 1935, even though Winston Churchill was aware of its black existence as early as October 1933.

While Göring delivered his speech, Heinreich looked around the auditorium. All the 250 officer cadets gathered there had graduated from technical hochscules with top notch grades, and all of them, like himself, excelled at athletics. They had been accepted to the Potsdam academy to become officers to serve in the various branches of the German military. Given his Castle Academy background in gliders, and his family connections, Heinreich had had no problems being accepted into the "commercial" pilot program. Thanks to his family's deep political connections, moreover, he had also known for quite sometime about the existence of the "black" air force. All Heinreich had to do was to get past the rigors of the Potsdam academy—something Frieda would not be able to help him do.

***

The *Kriegsschule* Potsdam military academy was a complete change of pace compared to anything Heinreich had ever experienced beforehand, and not at all for the better. 250 lost looking officer cadets from all over Germany were loaded onto several old buses and transported to the academy barracks several kilometers out from the town's rail station. The whole morning was spent signing paperwork, taking medical exams, filling in questionnaires, collecting new uniforms and boots and other sundry things. The entire afternoon was then spent back in the barracks, where groups of ten cadets were quartered in large rooms with ten drab bunks. To each group of ten officer cadets was assigned an unteroffizier who would lead the young men through the first six months of the program of the eighteen month long officer training program. Unteroffizier Beresel, who was portly, gruff and not in the least bit courteous, was assigned to Heinreich's group of ten.

Unteroffizier Beresel spent the rest of the day demonstrating to the new officer cadets the proper way to store their gear. "Uniform items will be hung in your closets with gaps of two centimeter spacing. No more than one millimeter variation will be tolerated. All buttons of stored clothing will be buttoned, all laces will be laced, and the beds, unless you are sleeping in them, will be made to military standards. Every morning, before our calisthenics, there will be a grooming and uniform inspection. No loose cables (threads) will be tolerated.

No lint will be tolerated. It is important that you officer cadets inspect and help each other. The word here is teamwork. I suggest after the dinner formation, you retire early. We will then begin the process of turning you ladies into officers and gentlemen Hitler can be proud of."

When Heinreich and his roommates returned to the barracks from their dinner formation, Heinreich tried his best to learn the names of his new companions. The name that stuck to him first was that of Sebastian Krall, who was also a city urchin hailing from Berlin. The next name that stuck to him was Peter Wohlthat, a farm boy from East Prussia. "Nice to meet you," Heinreich told them when it came time to choose bunkmates. "Shall we toss a coin Peter and Sebastian, to see who I bunk with?" Peter and Sebastian agreed, and when Heinreich caught the coin it was Peter that won the toss.

At 4:15 a.m. sharp, 250 sleepy officer cadets were rudely awoken from their slumbers. In Heinreich's barracks of ten, unteroffizier Beresel tossed an empty 55 gallon barrel into the center of the room. It came crashing down on the concrete floor with a resounding metallic boom, waking all but officer cadet Ebert Michaelis, who just snored on.

Beresel simply walked over to Michaelis' bunk, took hold of it with his two beefy hands, and tossed the sleeping cadet high into the air. Michaelis landed on his rump with a hard thud. This did wake him.

"Okay girls," Beresel barked, "now that I have your attention," he began, "we shall be off to an early ten kilometer run. Afterwards, you shall spend the remainder of the morning cleaning horse stables—"

A sleepy Michaelis, still rubbing his eyes, interrupted Beresel. "What!" the idiot exclaimed. "I came here to be an naval officer not to shovel horse manure."

Unteroffizier Beresel smiled at Michaelis. "I see," he said while walking over to the lank figure of officer cadet Michaelis. "For you, while the other officer cadets go to their breakfast formation, there will be an additional ten kilometer run. Any questions anyone?" While Michaelis stood with his mouth agape, no other cadet dared to speak a word.

After the ten kilometer run about a local obstacle course, the "ladies" were treated to a square breakfast. "Until you ladies become middle-class men, all your meals will be square meals," Beresel began. "That means that you ladies sit staring foreword with your heels placed together and your toes apart at a 45 degree angle. Your backs will be three inches removed from the backs of your chairs, and there will be absolutely no talking. Is this clear ladies?"

As not all ten of the officer cadets replied in unison, Beresel excused them from the breakfast table. "You will not eat until you can reply to me as a unit. Shall we go shovel some manure?"

This time the ten cadets managed to reply "Yes sir!" in near perfect unity. "That is better," Beresel declared. "We shall improve on it at the stables."

The morning session of horse manure shoveling went reasonably well, more so for those who had been around farm animals than for the city slickers.

"I guess," Heinreich joked to Wohlthat the farm boy, "before you can soar into the skies you have to shovel shit."

Wohlthat snickered. "At least, city boy, you haven't been doing it as long as I have. I'm here at this academy so that I never have to shovel shit again."

Heinreich agreed. "Point well taken farm boy."

Very early on, Heinreich and Wohlthat formed a fast friendship. Peter Wohlthat was as completely in love with flying as Heinreich was, and like Heinreich, he had also done much flying in one of the Hitler glider clubs. The two young men solidified that friendship by playing all manner of practical jokes on their fellow cadets, and even on Beresel. In their first coordinated joke they pulled off, they replaced the bullets for the first firing range exercise with blanks. The instructor of the range was none too happy when an entire row of fifty cadets, firing their first shot, missed. On another occasion they fed Beresel laxatives diluted in his soup when the squad was on field exercises far away from any commode. One joke that Peter was quite proud of was when he and Heinreich dug a six foot deep pit, filled it with mud mixed with horse manure, and then covered it over with dry dirt. That trap caught no less than the cadet commander himself. That prank cost the whole lower class a week of twenty kilometer runs with full packs after no one took responsibility for it.

Heinreich and Peter were also there for each other when the other needed help, like when Peter's father nearly died in an accident involving a tractor. Heinreich, through Frieda, had the injured man sent to Berlin to be treated at Dr. Reber's private hospital. The treatment Peter's father received there probably saved his life.

Lise's experience was a little more collegiate than Heinreich's. She burst into laughter when she received her first letter from her fiancée. It contained a photo of him, looking quite muscular without his shirt. He was standing outside a stable with one hand resting on a horse's rear end and the other holding a grimy shovel. *"The place stinks,"* he wrote.

"Sorry mein poor sweetheart," she wrote back. *"I think I've decided to go into nuclear physics. It is most fascinating. Now that I'm here at university, I know first hand that I am very fortunate.*

I am the only woman in the department. Despite my training at the Academy, if it weren't for my father's influence, I would not have secured a slot in the science college of this, or any other school. Hans, by the way, decided to join me here even though he literally had the pick of the world. He looks so young here, but that doesn't seem to bother him. He seems to be favoring mathematics over physics. Unfortunately, he continues to pursue his interest in Nazi politics. He attends all the rallies and has taken a position at the local office of information. I have to go for now—you know—classes and all. Of course it goes without saying, I miss you. I hope you like the picture."

The picture that Lise sent Heinreich was of herself and Hans standing outside the combined physics and chemistry building waving hello to him. The first thing Heinreich did with the picture was put it underneath his bunk's pillow. Every morning and every night he would look at it and would try to imagine what Lise was doing and how she would spend her day. It was hard being away from her, but like his father Karl had done before him, Heinreich felt deep within his bones that he was doing the right thing for Germany.

Hans did indeed stray away from physics, but only so far. He began to pursue the new developments in differential geometry being inspired by the nascent efforts to explain quantum mechanics in a higher dimensional setting consistent with the rules of Einstein's general theory of relativity. The boy had no notion that he was following in the footsteps of his very famous grandfather Albert. No—to him Albert Einstein, like the rest of the world's Jews, was a dirty Jew rat.

***

The months of the lower class portion of the Potsdam academy ticked by. From shoveling horse manure, the lower-classmen were graduated to marching, camping, survival training, small weapons qualifications, military history, and then more horse manure shoveling under the command of the much despised unteroffiziers.

Middle-classmen worked on drilling the lower-classmen and leading them during war exercises. As lower-classmen, Heinreich's group of 250 officer cadets were taught the enlisted side of the German military, while the middle- and upper-classmen, on the other hand, played the role of field officers and general officers.

While the unteroffiziers taught most of the lower-classmen classes, the middle-classmen and the upper-classmen, ran the academy as if it were a real, operational base. In turn, the middle-classmen took their orders from a real general officer and the upper-classmen. It was important for them, both lower, middle and upper class cadets, as the officers they would become, to know the world of the men they might likely lead in combat, and quite possibly to their deaths. It was a point well taken during several of the war game exercises. Even though they were only exercises, there were always mock causalities and deaths to deal with. As part of their training—and bonding—the lower-classmen then had to write letters to parents explaining how their compatriots had died. The middle-classmen, on the other hand, had to write letters to parents as to why their orders had led to the death of men under their charge.

The whole letter writing business served another useful purpose. It helped weed out those cadets who displayed manifestations of apprehension. "There is never any room for misapprehension in real combat whatsoever. Indecision gets your people killed," their commanders would tell the cadets after the exercises were analyzed and graded.

When the lower-classmen received their new dog-tags, the seriousness of the business they were in was driven home most poignantly. "Place one tag around your necks ladies," they were told. "Securely place they other one beneath the laces of your left boot. One day, that boot and that tag may be all that will remain of you to pass on to your dear mothers and sweethearts as mantle pieces in a jar of formaldehyde."

The military, the cadets learned, was more than just about fancy uniforms, pomp, circumstance and bravado. People could get killed being soldiers. It didn't help many of the cadets' moods that by now the German unemployment level was approaching one quarter of the working population, and that Hitler was increasingly talking up aggression. Heinreich, concentrating on what he had to do, rather than on things he could not change, kept cheerful. The pranks he continued to pull off with Peter helped.

Moods at the academy notably lifted as the Christmas break approached. The officer cadets would get to go home for nearly two weeks. The lower-classmen would then return as middle-classmen. It would now be their turn to play the role of field grade officers and lead the fresh lower-classmen into military life. The old middle-classmen, promoted to upper-classmen of general staff officer, would return to attend many a leadership, geography, and political science class for the period of a year. The graduating upper-classmen would be placed into real military service.

***

When Heinreich returned to Berlin that first Christmas break, Lise and he were inseparable. "Boy Heinreich," she said to him the first time she saw him disrobed. "That Potsdam school has done wonders for your physique."

Heinreich, always in shape, had put on several pounds of muscle, and his skin was bronzed from all the hours he had spent shirtless beneath the sun shoveling horse manure. "And university has been good to you," Heinreich replied stroking his hand through Lise's long hair. At the Castle Academy, the length of the girl's hair had been regulated to a certain length above the shoulders. At university, Lise had let her brown locks grow down below her shoulders for the first time in five years. "I love the new look," Heinreich would often repeat. When he and she made love, Heinreich loved to run his fingers through Lise's longer locks.

Hans was not a part of the reunion. He had returned down to Munich, to Bavaria actually, to attend several Nazi Christmas organizational meet-ups in preparation for the upcoming national elections.

"I would have liked to see the boy," Heinreich told Lise.

"I'm not so sure you would my sweetheart," Lise told Heinreich. "He has changed much, even much more than he had at the Academy. I worry for him. Currently, he is trying to form a committee to expel the so-called swing kids."

The news saddened Heinreich. "How is his mother? Is she divorced finally?" Lise told him that she was. "She's been divorced since September and is in Austria with relatives. Poor woman caught between two men at war—one of them her own son, and one of them her nasty ex-husband."

"We shall be nothing like that mein dearest," Heinreich declared. Lise agreed. She could never imagine her and Heinreich ever being so awful to each other.

Several times during that Christmas break, Lise and Heinreich discussed their future plans. The matter of Lise's education came up several times. "When I said you and I can get married after I become a leutnant, I wasn't thinking about you very well mein dear Lise. I figured you would follow me to flight school. Now I think that you should finish your schooling first. Women should have as many rights and opportunities as men. Besides, by the time you finish, I will nearly be an oberleutnant (captain), and be able to provide much better for you." Heinreich, as usual, always played down the fact that he was very wealthy baron. "Or I," Lise retorted, "shall be able to provide for you much better when I finish my doctorate dear baron von Onsager."

Back then in the 1930s the average time for earning a doctoral degree in the sciences was a mere three years after bachelors, and there was far more low-hanging fruit to pluck.

"Very well then Doctor Reber, we shall marry then in, say, 1933 or 1934. Is that a deal?"

"I think so," Lise replied.

As the Christmas break gave way to the New Year's celebrations, Heinreich and Lise took a brief trip to London to visit Frieda. Frieda was there visiting an earl with whom she had attended college. Being in England gave the young couple a nice change of pace as Frieda did little to chaperon the kids.

On New Year's Eve, dancing and celebrating at a popular club, Lise got drunk for the first time in her life. All too soon it seemed, the time came for Heinreich to return to Potsdam. The last thing Heinreich told Lise at the train station was that he thought life as a middle-classmen would be much better. "Now that I will have weekend privileges starting in February, you and I should be able to see each other at least thrice before I graduate as a brand new leutnant by year's end. I can't wait to earn mein own money and really be someone at long last."

Lise kissed her fiancée and pinched his cheeks "You silly boy, you are someone. You are my little sweetheart, as you shall always be."

***

Middle-classman life was anything at all but the privileged life Heinreich thought it would be. It was a nightmare from the start. Their middle-class responsibilities—shared with the unteroffiziers—were overwhelming. At first it was standardization—the task of getting 250 disparate lower-class personalities and backgrounds into lockstep—that stumped the middle-classmen. Heinreich was given cadet battalion rank, becoming a cadet major in charge of 50 lower-classmen and 10 middle-classmen.

While the unteroffiziers took care of teaching many of the lower-classmen classes as they had before, Heinreich and his fellow middle-classmen were tasked with teaching precision marching to the cadets in their charge. Within three weeks, Heinreich and the lower ranking cadet officers of his battalion succeeded in winning the first of the inter-battalion marching competitions. For their successful efforts, Heinreich awarded his exhausted lower-classmen with even more marching drills, and his battalion went on to beat out the four other battalions on every marching competition. To do so he had forced his lower-class cadets to march and drill even on Sundays. After all the lower-classmen of all the five battalions were marching well enough as a whole regiment, then began the first of the Potsdam war games, and even more stress for the middle-classmen now acting as field grade officers.

During this time Heinreich did not at all envy the middle-classman who had earned the rank of regimental cadet commander, cadet Oberst Eisenschmitt. Being in charge of a whole battalion was quiet enough for him. As it was, Heinreich barely had enough time to breathe a single breath in privacy. Even during weekends there were always so many things that needed to get done. It worried him, now that the war game exercises were beginning, that he might not ever have time to see Lise until graduation in December of that year. Fortunately, Heinreich's fears turned out to be unfounded. By the end of April, after the completion of the first round of war games, the lower-class cadets began forming into cohesive units. Heinreich had seen this happen before at the Castle Academy. Seeing it happen twice increased his confidence in the value of unrelenting drill and regimentation.

***

Hans paid a surprise visit to Potsdam in early May. He went there to attend an important political meeting concerning how the Nazis could best exploit the poor labor situation rampant in the area. Heinreich was surprised by how much the then nearly seventeen year old boy had grown. Hans, who had clipped his hair short in the style of Hitler and now wore the official black uniform of the SS, was almost a young man.

"How goes it old, privileged man," were the first words out of Hans' mouth when he entered Heinreich's dormitory.

Hans, Heinreich noted, seemed to be in very good spirits, and Heinreich, reaching out to take Hans' hand, laughed the remark off. "Not too badly you crazy wunderkind, and how is your dear mother?"

"She is doing wonderfully," Hans replied. "My father, however is not. He has had to shut down all but three factories. It saddens me about the workers he has had to let go, not of course the Jews. My father has retained only workers of demonstrable German descent."

Heinreich remained quiet. Personally, he didn't buy into anti-Semitism, or racism against any particular people, but given the times in Germany, he knew it was best to keep his opinions to himself. Many other people, too many others, were also keeping their mouths shut.

"I like your dormitory Cadet Major von Onsager. It is of good size. I see that, as usual, rank has its privileges, or is it the other way around? Privilege ranks."

"In some cases," Heinreich replied.

"I see you like to keep pictures of your fiancée everywhere," Hans continued as he started strolling about Heinreich's room. "I can personally vouchsafe that Lise is doing well at university. Every time I see her, she is sure to tell me that she misses you. And every time I see her, she shows me the latest pictures you have sent her. It is all very touching. When do you plan to marry her?"

Heinreich, trying to ignore Hans haughty affectation, liked the change of subject. "Would you like a shot of brandy Hans?" Heinreich asked.

"As for the brandy, thank you, but nein. By the way, when shall you two wed?" Hans repeated.

"Ah…we think we shall wed after she finishes her doctoral degree. I should be finished with flight school, and by that time nearly a oberstleutnant (captain). It feels like such a long while from now that sometimes I think I will not be able to endure it. But thinking about the sacrifices of my father helps."

Hans gave Heinreich a pat on his shoulder. "I'm glad for you. I knew years ago, perhaps before you yourself knew it, that you and Lise would end up together. What took the two of you so long to give in to your feelings I cannot say. As for your father, it is nice when people of power and privilege make sacrifices for their lesser beings."

That evening Heinreich treated Hans to dinner among the middle-classmen of the Potsdam academy and introduced Hans to Peter. "I tell you Peter, Hans is a genius," were Heinreich's first words to Peter while Hans and Peter shook hands.

Peter, noticing Hans' uniform, said, "So you must be an SS man. You should give them up and come join the real army!"

Peter was only joking, but Hans took it as a great affront. "You do know Herr Wohlthat, that the SS is Hitler's favorite, elite force? The army is cannon fodder to be sacrificed on the battle field for higher ends."

Peter burst out laughing. "Touché," he declared to the boy, erroneously thinking that Hans was jesting in the same spirit. Things only devolved from this point.

Dinner between the three young men that evening did not go well. By the following morning Hans was off to his meeting feeling glad to be away from all the athletic apes of the Potsdam academy. Returning to Berlin, he didn't bother to stop at the Potsdam academy to say goodbye. For his part, Heinreich was glad to have seen his old friend—though disappointed that Hans hadn't grown out of his Nazi craze.

For his twentieth birthday—which was still a few months away—Frieda bought her son a powerful, 1932 Deusenberg J LeBaron Phaeton, a popular touring car among American movie stars. "Pour le Baron von Onsager," Frieda had written in French in a birthday card placed on the beautiful machine's dashboard. The car—after Frieda had made arrangements with the base commander—appeared in front the Potsdam academy's main office.

Six weeks later, after her midterms were over, it was Lise's turn to visit Potsdam for an extended weekend. She was not very impressed by the academy and all its military mannerisms, still, the mannerisms were more "army" and less Nazi. Of course, Heinreich looked very spiffy in his cadet uniform. Heinreich, having long planned for Lise's arrival, delegated his command to his executive officer, Cadet Captain Wohlthat, and dashed off with Lise for the long weekend.

Needless to say, Lise was quite impressed with Heinreich's new automobile.

"When mein mother learned you were to visit me after your midterms several months ago," Heinreich explained after getting behind the wheel, "she wanted to make sure you and I had appropriate means of transportation."

"How sweet of her," Lise replied snuggling herself beside her fiancée.

"The car, I must say, nearly ruined mein reputation among mein fellow cadets. For a whole week I was called a rich boy. I didn't let this go on too far before fixing it. Regardless, I consider this car to be as much yours as it is mein."

Lise kissed Heinreich's cheek. "How fast does she go?" she asked. "Fast I assure you," Heinreich boasted as they drove off to visit Jena on the evening of that warm Thursday.

The town site of Jena was further south of Berlin than was Potsdam. It was where Heinreich—as Lise was fully aware—would attend Lufthansa's ostensible "airline" flight school program. Adolf Galland, who would become the most famous and dashing of the Luftwaffe aces of the second war world war, and the youngest general officer of the entire war, attended the program there in 1933 to earn his own "commercial" license.

For the first time in her life the notion that another world war might break out struck Lise as Heinreich pulled into the flight school. "Imagine if the French and the English learn about the true nature of this place. Might they not declare war on us?" she asked.

Heinreich, dizzy seeing all the high-performance aircraft buzzing overhead didn't see the danger. "You worry too much," he told her more than once. "The last war ended all great wars."

Lise liked the quaint town of Jena quite a bit more than

she did Potsdam. It had a university that had been founded in 1558, a very pretty Gothic town hall and an inn nearby, where she and Heinreich lodged a room for the weekend. "I could live here," she told Heinreich. "I would be your pretty little wife and tend to house while you flew around admiring me from overhead."

Heinreich chuckled. "Somehow, dear Lise, I don't quite see you as the type of woman who would stay at home. Perhaps you could become a professor of quantum, atomic physics...or whatever it is that you work on."

Though Lise very much enjoyed her time with Heinreich, he couldn't help noticing that something was troubling her during her visit. "What is the matter mein dear?" he asked her after they had finished making love, but Lise didn't answer his question. "I'm cold," was all she had cared to say. "Can you hold me in your arms and let me sleep in your embrace?" Heinreich kissed the tip of Lise's nose. "Of course mein dear."

It was during a private moment at the Potsdam rail station at the end of the weekend that Lise finally confessed she was losing her strength to be away from Heinreich. "I love what I am learning at university. However, I certainly do not need a university to continue with my investigations. Honestly, I don't think I can—or want to—hold out without you much longer. Perhaps we should wed after you earn your commission as we had planned originally."

Heinreich, who had been worried about Lise's attitude, was relieved, flattered, and even for a little moment between heartbeats, tempted to accept Lise's idea. "You must know it is very difficult for me to be apart from you as well, but I have another six months here before I become a real officer. We must remain apart until then. Why don't we discuss your education then, though I think I will still be of the opinion that you should finish your schooling. I want our children—our adopted children—to have a brainy mother."

It was Lise's turn to feel flattered, but also a bit alarmed. "You don't want me any more my Heinreich?"

Heinreich took Lise's hands into his own. "This is the silliest thing I've ever heard come out of your mouth dear Lise. You are mein life. You shall always be mein life." As Lise's' train huffed and puffed its way out the Potsdam rail station, Heinreich, looking on until it disappeared into the horizon, felt a pang in his heart.

***

After Heinreich earned his commission in the winter of 1932, he and Leutnant Wohlthat joined Lise in Berlin for a few weeks respite. Peter had never been to the "big city" and there was plenty of room at the ample von Onsager household. After having endured a year and a half of officer training, the two young men had become the best of friends. Lise had liked the country boy, with his blue eyes, golden blond hair, and affable smile since the first time she had met him. Peter Wohlthat was funny like Heinreich, and liked to talk up a storm about his childhood days on the farm. It gladdened her to know that Heinreich—who always looked out for other people—had a friend that looked out for him.

Frieda arranged a small banquet for her son the very first night the "boys" were in back in Berlin. They wore their new officer uniforms for the occasion—everyone wanted to see them in their uniforms. The Rebers, minus Olga and Nanerl who were left in the care of a nanny, came with Lise. She wore a beautiful evening gown made of gold thread for the occasion. Even Wolfgang, making the time to leave his busy ministerial duties in Munich, where the party headquarters of the Nazis were still located, managed to make it home to Berlin on time—literally as dinner was being served—using the special trains reserved for high ranking Nazi officials. "Being a minister," he both explained and apologized to his guests as the servants were putting out the first course, "has its advantages."

"Well boys, you did it!" Frieda began while Wolfgang took his seat at the head of the long table. Frieda raised her glass to offer a toast. "To my son Heinreich and his good friend Peter." Everyone around the table, raising their glasses to the air, uttered a cheerful here, here! in unison.

"I envy you boys," her cheery husband broke in. "You two are going off to fly into blue skies by day and starry, moonlit skies by night." He too offered the young men a toast of his own. "To our young men in uniform, to Germany, and my son, Leutnant von Onsager and his proud compatriot Leutnant Wohlthat." Another burst of smiles and here, here! erupted around the table. Minister Boltzmann was clearly very proud of his stepson, who was being given so much military responsibility. Heinreich was to be the student leader of the newly forming student squadron of "commercial airline pilots" he had been assigned to. In Wolfgang's eyes, Heinreich's achievements were demonstrating to the world the value of the accelerated training at his beloved Castle Academy. Heinreich was, after all, two years younger than Peter Wohlthat, who had also hailed from a good, but not an accelerated technical hochschule like the Castle Academy.

Seeing Heinreich in his new officer's uniform made Lise, who was sat between Heinreich and the baroness von Onsager, very proud of her fiancée, the more so when Heinreich paused the dinner celebration towards its end, when everyone had relaxed, to make an unexpected announcement.

Standing up Heinreich began, "Peter and I are, of course, most flattered and honored by this banquet, by the warmth that everyone has expressed for us. We are honored to be of service, but we would not have attained our achievements without the help of many peoples. I have to thank mein mother and Wolfgang for their trying years of looking after me, and teaching me well what the important things are. I also have a great debt of gratitude to someone sitting here at this table, Lise Reber. Since I met her in 1926, when she and I were but children, she has been by my side, aiding me, teaching me, urging me on to do better, and keeping me on the straight and narrow. Without her, I would have very little to live for." Lise, by this point in Heinreich's address was glowing brighter than her golden dress as Heinreich continued with his speech. "I wish to take this occasion to formerly announce mein engagement to the love of mein life, mein dearest Lise Reber."

Around the table everyone burst out in cheers and another round of happy applause as Lise, joining Heinreich, put her hands to his face for all to see her coruscating diamond, and kissed him deeply.

After the happy commotion and cheerful murmuring subsided, Heinreich, taking Lise's hand, continued. "We shall wed the minute—and not one minute longer—than Lise's graduation as Doctor Reber."

Lise, turning rouge, shed a few tears of joy as yet another round of applause and cheers broke out. When Heinreich in his turn kissed her again, it was the sweetest kiss she had ever had. If Heinreich had asked her to marry him there on the spot, she would have done so from where she stood. That she would have to wait was sad, but she could not doubt that Heinreich truly loved her.

"Dear Leutnant von Onsager," Lise replied after wiping a

sparkling tear from her cheek—everyone laughed when Peter handed Heinreich his handkerchief to pass on to Lise. "I shall finish my studies as rapidly as I can. I promise. I shall even make fleet-footed Mercury proud."

After everyone had retired that happy evening, Lise and Heinreich took a drive around Berlin in his Deusenberg. It was a pretty night, though quite cold at about 1 degrees Celsius. There were no clouds, the moon was out, and the city lights were burning brightly. The couple ended up in a café, talked into the wee hours of the night about past and future times, then drove back to her home not long before the crack of dawn. Along the way, Heinreich found a quite park. There, leaving his car running to keep the heater on, he made love to his fiancée with the full passion of the crystalline night and the joy of being once again reunited with his beloved.

At the entrance of her home, Heinreich kissed Lise goodnight. "I love you Lise," he told her.

"And I you," she replied.

"And we shan't be apart for very much longer any more," Heinreich continued. "Though Jena is farther, I am now an officer, and I will have the time to visit much more often. After I get qualified in the primary trainer, I have been told that I will even be allowed to fly it here to Berlin for the weekends, and you know me. I shall never pass up any opportunity to fly."

Lise, giggling, kissed Heinreich on the tip of his nose, and stepped into her house as he looked on with a smile. "Goodnight silly man," she told him before closing the door.

Heinreich drove off into the starry night feeling a sense of pure joy in his heart. He didn't have to be in Jena until another three weeks, and he planned on enjoying his time in Berlin to the absolute fullest. It would be the best Christmas break of his life, and, he promised himself, 1933 would be the best year of his life. He would finally be flying real aeroplanes.

Inside, Max, sitting in his favorite chair, was awake. When Lise entered, she noticed he had an old book parted open and resting on his lap. "What are you reading father?" she asked while taking a seat beside him.

"It's a book I used to read to you when you were newly born. Every night for months I would read the poem written in its first chapter. It seemed to soothe you and help you sleep when you were restless. I was reading it again to bring back old memories. I can still remember holding you in my arms as if it were yesterday. I cannot believe you've grown, but you have, and into a beautiful, bright woman. I'm happy for you—very happy. I like Heinreich. He is a good young man, and I know the two of you will be happy together."

For the second time that evening, Lise shed tears. "I love you father," she told Max while running her finger's through his hair. "I will always be your little girl." Max embraced his daughter. "I know," he replied.

***

Flight training was everything and more Heinreich had ever hoped it would be. The idea of being a Jungmärke (young pilot) pleased him to no end. The only problem that he had with the place was that it would be several months before he'd step into a real trainer. The months leading up to this point would be filled with ground school, aeronautics training, navigation training, parachute training, weather training, survival training, and so forth, to the tune of an abysmal pile of books that stacked up to his waist. Once again Heinreich heard the old Castle Academy's famous motto being repeated by his instructors. Being here, they would often repeat, is like taking a sip from an open fire hydrant. This time he was more than ready for the challenge. "Whatever the nature of the beast is, I will tame it," he would tell himself.

From the start, when Peter and Heinreich weren't busy training, they plunged into their books at every waking hour, often in a café near the airport where they could grab dinner and study with pots of coffee deep into the night. With Lise in his mind, Heinreich was determined to get things done right, and very appreciative he had only been sent as far away as Jena. Though he had known that Germany had been building a secret air force for quite some time, it was at Jena that he and the rest of the student pilots learned of the full extent of Germany's secret air force program.

Since 1923 Germany had been manufacturing military aircraft in the Soviet Union, and since 1925 German military pilots—restricted to wearing civilian clothes—had been training in secret air bases in Italy and in Lipetsk in the Soviet Union. Only in 1933 itself, after training over a thousand Luftwaffe pilots, was the Lipetsk air base and its training operations moved to Jena and other places within Germany. Heinreich would have hated being sent away to Russia. All that he had heard about the place was that it was cold, barren, backwards and ugly—and that the food was terrible.

"Would you have liked to have gone to Russia Peter?" Heinreich asked his friend the day they learned the truth about the "black" air force.

"Hell no!" had been Peter's prompt response. "Not even to bomb the place." Heinreich felt the same way.

"I guess mein friend that we got lucky to be here in '33, the first year the "black" air force program starts operations in Germany."

One advantage Heinreich had over his friend Peter in the

academics portion of flight school was that Heinreich, thanks to the endless tutoring sessions he had had with Lise and Hans, had learned as much higher mathematics as a typical mechanical or electrical engineer would be required to master. This advantage showed itself early on. Peter, as was typical of rest of the student pilots, had had only two years of calculus at his hochschule. Heinreich was thus able to anticipate much of what his instructors had to say, and even explain to them some of the principles behind optimized wing and propeller shapes. "You, Leutnant von Onsager," they would tell him, "should be a test pilot." Heinreich however, would always shrug his shoulders and utter, "Nein! Nein! Never. I shall be a fighter pilot like mein father."

If Heinreich was better at mathematics, Peter was the better student overall despite Heinreich's dogged determination. Whereas Heinreich had never developed good study skills despite Lise's best efforts to the contrary, Peter had perfected them to an art. Throughout his years in school, Peter had developed a system of using flashcards and mnemonics to memorize complicated things, a system that was most apt for an intense pilot training program. One of his mnemonics which became a favorite among both the students and instructors was GUMPS. It would go on into aviation history for decades to come. It stood for Gas/Gears, Undercarriage, Mixture, Prop, and Straps, and it was used during landing a plane to remember to lower the gears, turn on the gas pumps, set the mixture, flatten the prop and check the security of the safety straps. Peter made many such other mnemonics. Without Peter's flashcard and general help, Heinreich, with his duties as a student squadron leader, would not have fared so well. Together, the two of them, one better trained in higher mathematics, the other better at organized studying, formed a great team, and their squadron took an early lead against the performance measures of the other student squadrons.

During these months that Heinreich and Peter were running through their ground courses, Lise, working hard back in Berlin, at last selected her research topic. She was to determine if the were any observable symmetry violations in the production of electrons and anti-electrons in cloud chamber data. The universe was filled with matter. Where was all the anti-matter is what people wondered. It was an important question. As a tangent, Lise's research would require her to better understand nuclear transmutation reactions—the "chemistry" if you will of nuclear chain reactions—the very kind of reactions that would lead to atomic bombs, nuclear power plants and a cold, mad world in the not-too-distant future.

Hans also chose his dissertation project. He had a suspicion that the calculus of curved space in Einstein's general theory of relativity could be related to the "phase curvature" in quantum field theories. "If I am right, I am convinced that I will win the Nobel prize," he would often repeat to Lise. Unbeknownst to both Lise and Hans, however, was that they, along with a few other top notch graduate students throughout Germany, were all put on a collision course. Starting in 1935, a committee of Nazi scientists, including Hans' uncle Werner Heisenberg, would award a new scientific achievement prize annually to the best German dissertation project of a given year. The award, to be presented by Hitler himself to glorify Aryan science, would come with a prestigious professorship and a sizeable pile of grant monies.

***

Heinreich's first day of flying finally arrived in early April of 1933. As he walked up to the Heinkel He 63a, he thought it was the most beautiful thing he had ever seen. "She's more than a glider," Heinreich's instructor pilot uttered as he and his student began their walk around inspection. Major Rheitel, who had been a flier in the first world war, had taken a liking to Heinreich throughout the weeks of ground school. "Though I didn't fly with your father the baron, I heard good things about Karl. Let's see now how far the apple has fallen."

Heinreich grinned happily. "Let's," he replied as he clambered aboard the two-seater biplane.

By now Heinreich, along with all the other student pilots, knew everything there was to know about the trainer on paper. It was time to take their knowledge to the skies. From behind, Major Rheitel called out, "I will have my hands a millimeter from the controls. I want you to fly it off the ground Leutnant von Onsager. If you feel three shakes to the left of the control stick, that will be me taking over control of the plane. Is that clear?" Heinreich replied yes.

Heinreich was a natural. The He 63a lifted into the sky and climbed its way towards a bank of clouds about three thousand meters above. When Heinreich reached his designated altitude, Major Rheitel told him to execute a spin to the left. "After three rotations about the ground center point below, recover."

The purpose of spin training, in which a plane is made to spin nose downwards, was to give students confidence in their abilities to recover control from an aggravated situation. Heinreich complied by snap-rolling the plane into the spin configuration. Major Rheitel heard a "yahoo!" issuing from the young man ahead of him as the plane rolled inverted. Three rotations later, Heinreich snapped the wings level, broke the stall, added power and recovered the plane. "Excellent!" Major

Rheitel screamed out with a smile. Maneuver after maneuver it went like this—perfectly executed maneuver. The "kid" was a natural, just like his Castle Academy reports from Göring had said he was.

After Heinreich and Major Rheitel landed, however, the major put on his serious instructor's face. "It was not bad for a first flight," he told the young baron sounding as unimpressed as he could. The last thing the major wanted to do was to promote arrogant overconfidence.

Heinreich understood. "I will always endeavor to do better sir." Only one other student did as well, if not better, than Heinreich on his first flight, that being Leutnant Adolf Galland. Of course, Galland, slightly older than Heinreich, had already had some preliminary fighter pilot training in older aircraft in the Soviet Union during the months before the base at Jana was readied.

ALEX ALANIZ, PH.D.

PENULTIMATE WAR

If big changes came for Heinreich in 1933, bigger changes came to Germany. When the new national elections were scheduled for March of 1933, the Nazis, still only the second largest political party, were ready to pounce. They drew up lists of all their political enemies, including leaders of other parties, as well as of those who had publicly criticized the Nazis.

Mysteriously, the *Reichstag* (a central government building) was set afire. Hitler immediately claimed that the Communists were trying to overthrow the government and start a revolution as had happened in Russia. To prevent any such revolution, he pressed then President von Hindenburg to sign an emergency decree that granted Hitler and his party the right to take *any* measures deemed necessary to protect the state and cancel dangerous civil liberties. Von Hindenburg signed on February 1933, and just like that, Chancellor Hitler became the dictator of Germany. Any activity counter to the Nazi Party's liking was quashed without remorse.

When the March elections came, the Nazis gained only 44% of the votes. They were forced to form a government with the Social Democrats. Hitler's private army of Brownshirts and his private security forces the Blackshirts, however, began arresting political enemies in droves, sending the "traitors" by the thousand to Dachau, the first of the Nazi concentration camps. Thus, within a matter of months, the Nazis became the most potent political party in Germany.

Hitler's and the Nazis' war against the Jews also stepped up in pace during this period. Whereas before, anti-Semitic rhetoric had helped the Nazis win votes, the Nazis now had the power to put some of their vile notions into play. In April of 1933, Jews were banished from government jobs and a stringent quota was established banning them from attending universities or professional schools. A general boycott of Jewish shops was also enacted.

During the period starting from von Hindenburg's signature of Hitler's decree, the national media, radio and newspapers alike, fell under the control of the Nazis. The first of many books designed to promote Nazi ideology began to appear. Thus, in the matter of mere months in which Hitler had consolidated his powers, the whole of Germany underwent a massive change of state. By the end of 1933, the old Germany was gone.

***

Heinreich welcomed the end of 1933. With his primary aviator's training completed by the end of early November, he looked forward to advanced fighter training, and to a much needed, well-deserved break with Lise. He said his goodbyes to Peter for the interim, and dashed off to Berlin in his personally assigned Heinkel 63a.

"Lise," Heinreich told her at the airport, "there is no way you aren't going to fly with me now that I've earned my basic license."

"I look forward to it," Lise replied after kissing her pilot.

Lise, of course, was quite excited to be reunited with her fiancée. She figured she was no more than two years away from finishing her dissertation work, and was beginning to see the light at the end of the tunnel. Her primary research project was turning out negative results. Her related work on nuclear transmutations, however, was yielding fruit.

As usual when Heinreich would return home there would be the obligatory family dinner between the von Onsagers and the Rebers—this time it would be at the Rebers' household. Wolfgang, being too occupied with his duties, could not make it.

Heinreich was surprised to see how much Olga and Nanerl had grown. Olga, still devilishly playful, was already six years old and Nanerl, now three years old, was talking up a storm. "Do you see the giants in their cloud houses when you fly?" Nanerl asked Heinreich when she first saw him. "Why yes," he replied while bending over to pick her up. "And I see giant dragons that fly over their castles." Nanerl giggled as Heinreich raised her high into the air above his head. "Do you see the dragons?" he asked her. "No!" she replied.

Frieda, seeing her son playing with the child, felt a tinge of sadness that Lise could never give her boy a son or daughter of his own. Lise, also watching the happy scene, felt the same tinge of sadness, but she understood that Heinreich loved her, and that they would adopt children—many, many little pilots as Heinreich would often tell her. "Five or six," he would say, and she feared he meant it.

This time for the New Year's celebrations, Heinreich and Lise traveled to Italy to welcome in 1934. They stayed in a beautiful *pensione* by Lake Como north of Milan, a jewel-like oasis of tranquility amid snowy alpine peaks and a lush covering of Mediterranean foliage flowing between ancient Roman villas and 400 year old houses. It was a perfect place for a young couple in love.

Clasping hands during a morning walk along the lake by decrepit Roman columns, Lise told Heinreich that she was very concerned about Hans. "He seems to be trying to race against me all of the time now. If he is not talking about the Nazi world that will arise, he is telling me he will finish his dissertation before I finish. He even believes he will earn a Nobel prize."

Heinreich was a little disturbed to hear the news. "How old is he now?" he asked.

Lise had to think for a second. "Well he was nine in '26, so he must be about seventeen now."

Heinreich sighed. "Well perhaps that explains things. The boy is full of hormones. He needs to find a sweetheart and think less about physics."

Lise considered Heinreich's remarks for a bit as they walked on past the ancient columns. "Perhaps you are right," she concluded. "I will try to remember this the next time I see him and he tries to offend me."

While the couple walked on hand in hand along the water's edge, a cold breeze began blowing across the lake. "I am getting a little cold Lise, perhaps we should stop somewhere for breakfast." Lise agreed. She too was feeling the chill of the winter air flowing down from the glaciers located nearby at the Swiss border. "If we are to go skiing, I think we should buy supplies after breakfast."

Turning around and walking back the three kilometers into town, Heinreich and Lise found a quiet café at which to breakfast. After the coffees were served, Lise asked Heinreich about something that had been bothering her for quiet some time. "I know dear Heinreich that you don't fear flying, and I trust you are an excellent aviator, but do you ever fear war?"

Heinreich had not expected such a question. "Why do you ask such a question Lise?" At the moment Lise was about to reply, the waiter came by to take the couples' order. After he left, Lise continued. "I ask because...because I am being silly. I want us to have our own home, some children who need parents, and a long, long life together. Sometimes I think you are following too closely in your father's footsteps, and the Nazis are very warlike."

Heinreich took Lise's hand. "Look," he replied, "I think the Nazis are a little insane at times, but their goal is to make Germany strong. We have been so defenseless for so long, and our peoples—they say a full quarter of them—are not working. I agree with some of their ideas. I don't think they want more than to buttress Germany's defenses."

Lise considered Heinreich's response between a sip of coffee. "I hope you are right, but these times we are in do frighten me."

Heinreich smiled and reached over to kiss Lise's cheek. "Oh Lise, I think you are worrying about things that are crazy and cannot do anything about."

Lise wasn't as sure as Heinreich. "We can leave Germany as my father has thought about."

Heinreich snickered. "What? Your father leaving Berlin? It will never happen. He won't take the family out of Berlin, much less out of Germany. Your family's whole life is planted in Berlin." On this, Lise was afraid that Heinreich was correct.

Later that morning, while Heinreich and Lise went out to buy food and drink for their ski trip, they passed by a boutique for wedding dresses. A particular dress caught Lise's attention. "Heinreich, would you mind if we go into the boutique?" she asked. "Of course not," Heinreich replied.

Taking Heinreich by the hand, Lise dragged her fiancée into the old boutique and right up to the dress that had caught her eye. "It's lovely," was all Heinreich could say. "So you think I would look beautiful in something like this?" Lise pursued.

"Of course, and I mean it dear Lise. You would look beautiful to me no matter how you dressed, or didn't." That comment got Heinreich a naughty, naughty wave of Lise's finger and a smile from the boutique's owner who had approached the couple at that instant.

"Are we planning to get married soon?" the old woman asked in a broken German.

"As soon as I finish university," Lise replied. "And not a second later," Heinreich added.

Lise hoped to have her dissertation completed and defended by the end of the year, and to get married no later than January of following year, that being the January of 1935. She was hoping to have her wedding take place on the island of Capri situated off the coast in the Bay of Naples. The island, which Heinreich and Lise had toured about a year before, jutted high out of the sea with sheer walls. High atop the easternmost crag of the mountainous island laid the ruins of the largest of the twelve villas built by Emperor Tiberius. Up there in the ruins, the vista of the surrounding ocean, the waves crashing against the rocks, the fishermen's boats bobbing in the tiny seaport a thousand meters below, and the village homes with their colorful flowerbed gardens nestled along the slopes was spectacular. Heinreich, however, still hadn't changed his mind about holding the wedding in Paris, in the Cathedral of Notre Dame as he had first wanted. This was a detail the couple needed to work out and enjoyed debating over when they would lie in bed together.

***

1934 passed by faster than Lise and Heinreich had ever imagined possible. Before they knew it, each had collected a pile of post cards, photographs and letters from the other running from January through early December.

Lise managed to defend her dissertation by March of 1935. She just missed out by a matter of weeks on a discovery that would have earned her the Nobel Prize. Instead James Chadwick, who had been a prisoner of war of the Germans in first world war, beat her to the discovery of the neutron. When Lise read his paper, all she could do was laugh. The data she had collected, and that hinted at the neutron, had been in front of her nose, she just hadn't recognized it. Hans also managed to beat Lise as he had promised he would. He defended his dissertation several months earlier, finishing his doctoral research in late 1934 as a seventeen year old prodigy.

While Lise finished things up in Berlin, Heinreich and Peter Wohlthat were kept extremely busy flying the more advanced Heinkel He 66, a sleeker, faster, more powerful single wing prototype fighter. It was far more agile than the primary trainers they had flown the year before at their primary pilot training school. During the months of 1934, the pilots of Heinreich's group mastered formation flying, night flying, dive bombing and close air support tactics, and, best of all, air-to-air combat, as in dog fighting. Pulling gees, yanking and banking, hunting and killing was a blast. The only real dark moments came when two pilots crashed into each other, resulting in their immediate deaths, and when another pilot killed himself only a week later during a botched landing.

The wedding between Lise and Heinreich, as he was ordered to go to Italy for secret training operations with the Italian air force in February, suddenly had to be delayed until August of 1935. Fortunately, with so much advanced notice, both Minister Boltzmann and the Baroness von Onsager would be available.

Between Paris and Capri, Lise won out. The wedding, Lise told Heinreich, would be held in Capri amid the ruins of

Tiberius' villa. "It's the least you can do you scoundrel," she told Heinreich, "after delaying our wedding yet again." Heinreich, feeling guilty, graciously accepted Lise's ultimatum. He saw to it that a bouquet of roses were sent to her laboratory each week.

Though Heinreich wasn't supposed to, just before flying off to Italy, he whispered a secret into Lise's ear. "The existence of the Luftwaffe will not become official until May of '35. The good thing about this is that I, along with all the rest of Germany's "commercial airline" aviators, will be coming out of the "black" air force and into the open light. There will be no more sneaking around."

"That's a relief," Lise whispered back. "I think our lives will be much better." She hadn't liked all the secrecy and sneaking around in Heinreich's life, never knowing when he might have to go to some secret training operation.

***

While political upheaval roiled in Germany, the von Onsagers and the Rebers enjoyed the remoteness and tranquility of Capri. Both families looked foreword to seeing Lise and Heinreich finally wed, and it was a relief to get away from the ever present ugliness, darkness and hatred back home.

The Rebers were the first to arrive on the rock island. Frieda and Wolfgang arrived the following day. Heinreich, busy finishing up the last of his duties in Italy, would come by the end of the week. Lise, of course, was nervous. She had no doubts about Heinreich showing up, but the fact that he wasn't there made her have silly thoughts of being left at the alter. She, together with her mother and Frieda, occupied herself with the innumerable details of the wedding. Even though it wasn't going to be a very large affair, it was still a wedding.

Heinreich, back at his secret Italian air base, was chomping at the bit to get to Capri. If it wasn't for Peter keeping him grounded, he would have gone mad during the last few weeks of his bachelorhood. When Heinreich least expected it, on the day before his departure, just as he was climbing out of his airplane from a mission of simulated dog fighting, a gaggle of fifty other German pilots, with Peter in the lead, accosted him. "Before we let this rooster have his wings clipped," Peter announced to the group of aviators that had encircled Heinreich, "we should make sure he has a chance to fly the coop." All the pilots around him burst out in cheers. "To the club!" they all began shouting. Despite Heinreich's protests that he had to prepare for his departure, he was dragged off to a local bar the pilots favored—they had reserved the place for the entire evening.

Prostitutes, of course, showed up to the bacchanal by the dozen, and as the evening progressed and the beer flowed, Heinreich's party got increasingly rowdy. A round of spontaneous "crash landings" began. Four pilots would grab a fifth pilot by his limbs and, with a running start, toss the "plane" onto the bar, sending the tossed pilot sailing across the slick, beer-soaked surface. On the other end of the bar were the "stopping bumpers" as provided by some prostitute's bare bosom. For each landing the lady received so many lira and kisses on her nipples. There was no lack of volunteers for these crash landings. Not even the senior instructor pilots—many with wives and children—nor even the squadron commander missed out on their turns.

The following morning Heinreich awoke at the table where he had last sat. Beside him, laying in a pool of his own cold vomit slept Peter. Heinreich had a pounding headache and not much of a memory of what had occurred the night before. His shirt was torn open, but his trousers, he noticed, were still on. He took that as a good sign. When Heinreich looked at his watch, he nearly passed out. He had missed the express to Rome. It would now be impossible for him to make his way to Naples on time to catch the ferry to Capri for his wedding the following morning.

When Heinreich tried to wake Peter, Peter wouldn't budge. Heinreich had no choice but to fill a tall glass with beer and splash it over Peter's face. It helped wash off the larger chunks which clung to Peter's cheeks and hair. When Peter's eyes finally opened, Heinreich put his watch in front of Peter's face. "Do you see the time Peter?" he asked his friend.

"My friend…your goose is cooked," Peter replied before laying his head back down upon the surface of the soiled table.

"Nein, nein. It is your goose that is cooked if you don't find a way to get me to Capri on time."

Peter hardly stirred. "Don't worry. Don't worry. I will fly you there."

Heinreich didn't see this as a solution. "It won't work. Our Italian unit is still a secret remember? And Capri has no landing strip."

Peter managed to raise his head. "You will parachute you idiot. Now please stop yelling at me," he pleaded while wiping a piece of partially digested sausage from the corner of his mouth. "Now pass me that bottle of beer. I need to wash my mouth out."

Two hours later, after having stumbled into a two seater fighter, Peter and Heinreich were off, showered, cleaned up and in fresh uniforms, but still somehow wreaking of cigar smoke, beer and vomit. Fortunately the sergeant in charge of the base kitchen, having anticipated the effects of Heinreich's bachelor party, had made plenty of strong, black coffee, and put some in a large thermos bottle for the young men. "Good luck," he told the two green looking aviators, "you knuckleheads will need it. And don't forget to refuel, Naples is far away."

In his haste to leave for Capri, Heinreich forgot to wire Lise that he would be coming by "alternate" means. Naturally, when Heinreich didn't arrive at the appointed time at Capri's seaport, Lise panicked. Her worst fears were confirmed.

When Minister Boltzmann—going over the mayor's office where the only phone on Capri was located—called the secret air base in Northern Italy, all he was told was that no one had any idea as to where Leutnant von Onsager might be. Wolfgang didn't know what to do to reassure the Rebers that his stepson would show up except to babble. "Heinreich is a good boy. He will show up."

Lise, nearly inconsolable, was beside herself with grief and worry as the hours passed by. "Don't worry dear. I know Heinreich must be on his way," Frieda would add to Wolfgang's words before telling her husband to call the base again.

More hours passed. When Wolfgang phoned the air base for the tenth time, he received a most worrisome message. "No Minister Boltzmann, we have no notion where your son may be," Hauptman (Captain) Molders informed Wolfgang. "However, we have sent troops to check the local morgues and police stations."

"Thank you Herr Hauptman Molders," was all Wolfgang could say when he hung up feeling numb.

When Hauptman Molders hung up the phone, he and a dozen other pilots burst out laughing. Heinreich's friends were fighter pilots being fighter pilots. They knew, as Peter and Heinreich had called from their refueling base, exactly what was going on and were just having a little fun at the minister's expense.

A 6 o'clock the sound of an airplane's engine was heard circling over the island of Capri. As the sight of airplanes was still rare on that island, scores of children came rushing out to see the plane circling about, descending down ever lower with each orbit. A worker from the hotel came knocking on Minister Boltzmann's door. "Sir, there is a German airplane circling about." Wolfgang grabbed Frieda and then the Rebers. Everyone went outside to see the spectacle.

"I don't know where the hell they are Peter," Heinreich called out to his friend over the intercom. With his binoculars he could see the ruins of Tiberius, but he wasn't familiar enough with the island to remember the palazzo where the members of his wedding party were staying. "Maybe it's that one, the big one with the large gardens."

Peter took Heinreich's cue and dived down in perfect dive bombing configuration towards the presumed palazzo. The whine of the plane's engine revved up as the plane gained speed.

The lower the plane got, the more nervous the people below got. Peter began calling out altitudes. "350 meters, 300 meters, 250 meters, is it the place Heinreich?"

A second later Heinreich replied. "It is!"

At a mere 150 meters above the palazzo, Peter pulled up hard as the people below began running for cover. "Are your straps on?" Peter called out.

"Yes Peter. I will jump when you reach 1000 meters, any higher and the damned winds may blow me out to sea."

Peter opened the canopy. "Don't worry about me Heinreich. I will land in some farm somewhere near Naples. I will get to Capri by morning if not by tonight. I wouldn't miss being your best man for the world."

At the appointed altitude, Heinreich stepped out onto

a wing. "I hope I never do this again," he called out. Peter, placing his hand flat against Heinreich's face, shoved the groom out of the plane. "Get out of my plane," he told Heinreich as he did so.

Down below people gasped and began to murmur frightened things. The servant next to Lise began exclaiming in Italian, "The poor pilot has fallen out of the plane!" For a second Lise became convinced that whoever was up there had indeed fallen out the plane as the frightened Italian had thought, but then the plane rolled away with a hard snap roll. The pilot was clearly still aboard.

At 500 meters, tumbling earthward at nearly two-hundred kilometers per hour, Heinreich spotted Lise. He swallowed hard. He had little doubt that he was in trouble as he pulled his chute's rip cord. To everyone's relief down below, the parachute of the person plunging earthward above them, popped open and filled with air. At 100 meters above the palazzo, Lise recognized the figure that was waving at her and her blood boiled. A few seconds later Heinreich landed in a muddy flower bed and tripped over a rock. Only now, as Heinreich removed his goggles and shook some of the mud from his body, did Frieda, Wolfgang, Max and Belinda begin to recognize the man in the parachute.

"You stupid, stupid man!" were Lise's first words when she reached her fiancée. "You had us all worried you were dead. Do you know they are looking for you in the morgues?"

Heinreich had no idea as to what Lise was talking about.

Lise then burst out into tears. "I'll shall never forgive you for this," she told Heinreich as she struck him on his chest twice. But then, taking a good look at her fiancée as he grabbed her by her shoulders, with his mud-smeared face— except for the clear ovals around his eyes where his goggles had been—his wind blown hair, and his look of utter cluelessness, she began to laugh.

"I'm sorry," were Heinreich's first sheepish words.

Lise's laughter, mixing with her pouring tears only confused Heinreich as the Rebers and his parents walked up to the trampled flower bed. Before long, even Heinreich, who wasn't quite sure why, began to laugh, especially when the manager of the palazzo walked up to Lise and asked her who the gentleman in the parachute was. Needless to say, the event became the talk of the island that evening and well into the next day when many of the islanders decided to attend the wedding at the ruins.

Peter, as he had promised, managed to make it to the wedding on time. He managed to hitch a ride on a fisherman's boat from Naples to the island later in the evening. While the mothers of the bride and groom, and many of the nearby womenfolk shed tears, and a horde of islanders crouched in closer to get a better look at the crazy man who had fallen out the sky the day before, Lise and Heinreich floated high in a world of their own.

"With this ring, I, Baron Heinreich von Onsager do thee wed," Heinreich spoke as he placed Lise's ring on her trembling finger. Lise followed in her turn. "With this ring, I, Lise Reber do thee wed." The old priest than spoke his final words with a smile, concluding with the expected, "I now pronounce you man and wife." A wave of cheers broke out and swept down the island as Heinreich kissed Lise for a good, sweet while.

***

The wedding of Heinreich and Lise occurred on 1 August 1935. On 2 August 1935 President von Hindenburg died. No new president was appointed. Instead the powers of the chancellor and president were combined. It was a bad year for the Jews of Germany—the year of the Nuremberg laws.

The Jews had suffered through various boycotts since as far back as 1933. In May of 1935, the year the existence of the Luftwaffe became official, Jews were forbidden to join the Wehrmacht (army) and anti-Jewish propaganda began appearing in Nazi-German shops and restaurants throughout the land. On 15 September 1935, the "Law for the Protection of German Blood and Honor" was passed preventing marriage between any Jew and any non-Jew. At the same time the "Reich Citizenship Law" was passed. It stated that all Jews, even quarter- and half-Jews, were no longer citizens of their own country. Jews became merely guests.

In December of 1935, after two Jewish dissertation competitors had been disqualified from participating in Hitler's first national science award, Lise was proclaimed as the victor. The dictator personally presented the award to her before a large crowd gathered at the university soccer stadium in Berlin before proceeding with his planned speech. "That Castle Academy," he announced to the crowd after giving Lise her medal, "is worth its weight in gold. And our Doctor Reber deserves our appreciation for her fundamental work in physics. I have no doubt that with her award monies she will go on to contributing to German science."

It boiled Hans' blood that Lise had beaten him, and he made a point of not making it to the rally. "One day," he promised himself, "Lise's fortune will run out."

***

The next time Heinreich and Lise saw Hans was in the spring of 1936. They showed up to congratulate him for his promotion at his new SS office.

"I am happy for you Baron von Onsager and Doctor von Onsager," Hans addressed Heinreich and Lise when they walked into his new office.

"Why so formal dear Hans?" Lise asked him.

"Because, dear baroness, I understand how things work. Because Minister Wolfgang is Heinreich's father, and his mother the baroness cavorts with the wives of all the Reich ministers, I have little wonder why you Lise won the first Hitler science award."

Heinreich and Lise were flabbergasted.

"I assure you," Heinreich replied, "that neither mein step father nor mein mother had anything to do with Lise's award. You know better than I that her work is important, and that she barely missed out on the discovery of the neutron."

Hans turned beet red. "And my work? I'll tell you about my work! Someday people will recognize it as the greatest work of this century, if not this millennium. I will show people can leap across space using the very "spooky" physics the Jew-rat Einstein abhors. Now take your leave."

Lise tried to approach Hans only to get yelled at by him.

"I said do please take your leave!"

Heinreich took his wife's hand and held her back. "Come on mein sweetheart. We are clearly not welcome here."

Whereas Heinreich took offense, Lise was saddened by the unexpected turn of events. "He is young and passionate," she told her husband as they walked away from Hans' office. "He will come around."

Heinreich disagreed. "I think he is gone."

***

Not long after the award ceremony, Hitler took a big gamble and sent troops to the Rhineland in March, a demilitarized zone of the Versailles Treaty. About that time a massive civil war broke out in Spain during the summer of 1936.

On 25 July, Hitler and Benito Mussolini agreed to support General Francisco Franco and his nationalists against the republicans, which consisted of the Communists, Socialists, and Basque and Catalonian separatists. Italy sent General Franco 700 planes. While the United States and Britain decided to stay neutral, many volunteers from both nations shipped off to Spain, including a group of American pilots who helped form the Republican Air Force, composed of a hodgepodge of airplane types and pilots of different nationalities. The Soviet Union was also quick to offer aid to the communists, and France provided the communists more airplanes and some artillery. The Luftwaffe quickly organized the German Condor Legion, and equipped the Legion with Germany's most modern aircraft, the Heinkel He 111, a two engine bomber, and Wilhelm Messerschmitt's Bf 109 Gustav, a fast, agile fighter.

Heinreich saw the Spanish Civil War as a great opportunity to advance his military career. Unfortunately, he was denied his request to join the Condor Legion. Instead, while Peter was accepted, Heinreich, because of his excellent piloting and mathematical abilities, was promoted to hauptman (captain) and sent off to become a test pilot for the Messerschmitt company at the Augsburg's military facility northeast of Munich, where a sprawling, extensive network of subterranean, steel-reinforced, concrete-walled rooms, hallways, and doorways had been dug out. Heinreich, longing to fly combat since he could remember, hated the place.

Lise was relieved. Augsburg, which had been founded by the Roman Emperor Augustus in 15 B.C. at the junction of the Wertach and Lech rivers, was small, quiet and far removed from the bullets and bombs flying in Spain.

1937 brought more turmoil to Spain. The American Lincoln Battalion moved to the front lines in February. On the 17th of that month, the battalion suffered its first casualty, Charles Edwards, and Leutnant Wohlthat quickly became an ace with his first five kills. By July the Lincolns had suffered so many casualties, they were forced to merge with the Washington Battalion.

By April 1 of 1938 only 120 of the original 500 Lincolns remained after being overrun near Gandesa. Peter Wohlthat had by then collected twenty-two kills on his belt. In Germany, meanwhile, fully one quarter of the Jewish population had left the country. On 1 March 1938 Jewish businesses could no longer be awarded German contracts, and the pace of beatings and killings of the remainder of Germany's Jews accelerated. Hitler annexed Austria (with 183,000 Jews) to Germany later that month. During the Evian conference in Evian France later that year, a conference which President Franklin Delano Roosevelt had called for to address the refugee question, delegate after delegate of the attending nations expressed sympathy for the plight of the German Jews, but did not raise their extremely smallish quotas for Jewish immigration.

Hitler, responding with great pleasure after the conference, stated, "I find it astounding that foreign countries which have criticized Germany for our treatment of the Jews, to not want to open the doors to them [the Jews] when the opportunity was offered. We, on our part, are ready to put all these criminals [Jews] at the disposal of these countries, for all I care, on luxury ships."

The writing was on the wall and Hitler read it. The world did not care about the fate of the Jews. On 17 August 1938, all male Jews had to add the name Israel to their names. All female Jews had to add Sarah to their names. On 30 September 1938, it was decreed that Aryan doctors could only treat Aryan patients, and as Jews were banned from being doctors or holding professional positions, there was little Jews could do for medical treatment.

Dr. Reber, much to his regret, lost a fifth of his patients for being Jews. He did what he could by having "guests" visit his house, but he had to exercise extreme caution not to put his own wife and daughters in too much danger. One day Max uncharacteristically smashed a pastry saucer to bits against a wall when he learned a former and dear patient of his, a ten year old boy he had delivered, had died for want of a simple surgery. "Why did I bother to help bring Marc into the world?" he asked Belinda. "You can't blame yourself for the evil of others," she tried to console him. "God knows you do as much you as can."

Hans thrived during this period. To his delight, he was given a powerful position in Hitler's SS political unit under Goebbels' command. "Beginning at the universities, my men and I will find as many Jews as I can," he told his new boss. He proceeded to do so with much aplomb.

During October of 1938 Polish Jews were herded like cattle and dumped at the Polish border. On 7 November 1938, under the advice of Goebbels and the wunderkind Hans Fritz, the SS conducted the Night of Broken Glass (Kristallnacht) in which storefronts of Jewish shops and offices were smashed. 100 Jews were killed and 20,000 were sent to the newly formed concentration camps. Hans himself—getting warm blood splashed on his face—shot two Jews dead for refusing to obey his orders. "I told the father rat to follow me," Hans boasted to his men. "He called me a boyish looking thug.

Before he could utter another sentence, I relieved him and his son of their brains with one shot apiece."

Exactly a year after the American Lincolns were overrun, General Franco declared the war ended on 1 April 1939, and Germany's Wehrmacht and Luftwaffe forces returned home with much experience. Neither Lise nor the Baroness von Onsager had any hopes for peace. Their beloved Heinreich would likely go to war. Heinreich, however, seeing himself stuck as a test pilot, doubted with great frustration that he would ever see any action.

***

War came to the world in 1939. British Prime Minister Chamberlain's sacrifice of Czechoslovakia to Germany on 15 March 1939 failed to appease Hitler's war lust. The concession of the world, rather, emboldened him. That same day German troops crossed the borders of Czechoslovakia, and Czechoslovakia ceased to exist as a country.

In late August of 1939, Hitler escalated things. SS troops took twelve prisoners out of Buchenwald, forced them to take poison, and shot them after they had put on Polish uniforms. An SS Officer yelled in Polish into a radio that they had come to invade Germany, and then the SS fled. Thus on 1 September 1939, Hitler sent troops into Poland, and a stunned world learned the meaning of the word blitzkrieg as tanks rolled into the hapless nation. Two days later, only now realizing the error in their ways, both France and Britain declared war on Germany. On 17 September, the Soviets, invading western Poland, entered the fray. Thus began World War II.

By 1 October, Poland was conquered. Hans, having dedicated himself to his career in the SS, had essentially given up working on his strange theories of phased spin spacetime.

He was assigned to a special unit to help steal the science and technology of vanquished nations, as well as to continue his old line of work of tracking down Jewish academics to be shipped off to concentration camps. Hans' first assignment was Poland. He departed with celerity.

One week after the invasion of Poland, to his orgasmic delight, Hans received the greatest gift of his twisted life. The German Army captured a young professor—Emil Goldstein— among a camp of displaced refugees on the outskirts of Warsaw. The twenty-six year old professor, who had been born and raised in Bonn, was turned over to the SS, whereupon he was thrown into a local jail with other persons of interest. A young SS officer told Emil that it would be in his best interest to betray other Jewish scientists in hiding.

Emil held out well for three days under relatively mild questioning until Hans arrived in town and took over the interrogation of the prisoners. "Deny the bastards water," he promptly ordered his men. "In three days I shall return. I promise you we will break a few of them before lunch." Hans was in a cheerful mood.

On the appointed morning, Emil was the first of the prisoners taken to a makeshift interrogation chamber. Without uttering a word, as per Hans' instructions, Sergeant Schwartz, built short and stout like a barrel, began beating the young professor's face while Hans took a seat in the background and awaited his breakfast. Schwartz kept the beating up for about five minutes, breaking off several of Emil's front teeth and bloodying his face. With his hands bound behind his back, Emil, who wanted to start talking after the first minute of his beating, was struck hard in the mouth each time he tried to utter a syllable. Twice he swallowed a tooth.

"Halt!" Hans eventually cried out. "Now on to his fingers," he ordered Schwartz.

"Happy to sir," Schwartz replied. "Have his hands unbound," he ordered Corporal Bailin. An SS private entered the room at that moment with a pot of coffee, juice and Hans' egg and ham breakfast. "Thank you Greiner," Hans told the teenaged private.

Before Emil could realize what was going on, Sergeant Schwartz had grabbed his right hand. Slowly, so as to inflict the maximum pain, he pulled Emil's index finger back. Emil began to moan and kick under the excruciating pain. It was useless to try to wrestle his arm free from Corporal Bailin. The more pressure Schwartz applied, the greater the pain until, pop, out came the finger bone from its joint. Schwartz then repeated the procedure twice more. Hans, sitting at a table in the background, enjoyed his breakfast as he looked on, taking sips of his coffee between Emil's screams of agony.

Before Schwartz proceeded on to the next finger, Hans stopped him and walked up to the sobbing man. "Professor Goldstein, it hurts?" he asked Emil. "All you have to do is give us one name, and all of this will stop. I promise."

Emil, dehydrated, his face a bloody pulp, barely able to see through the slits that had become his eyes, relented. "Please no more...I will...will help you," he told Hans. "Just of sip of water first, please sir...so that I may talk well."

"Why certainly professor. Sergeant, be so kind as to get the good man some water," Hans told Schwartz. "And bring me my notebook while you're at it. The professor wishes to sing for us." A minute later Hans had his notebook out while Emil, dripping bloody saliva, tried to drink through his battered mouth.

"Well Professor Goldstein, what would you like to tell us today?"

Emil coughed out several wads of semi-coagulated

blood. "My father...he was born in Bonn...told me he was childhood friends with..." Emil took another sip of water before continuing. "...friends with the mother of Professor Lise Reber, the lady who was awarded Hitler's science award."

Hans eyes perked up. "You mean Professor Lise von Onsager?"

"Yes," Emil replied, "she."

"And what of her mother?" Hans pursued.

"Her mother...she is a Jew," Emil said gagging on his blood.

Hans' heart skipped a beat. "You are sure of this? Professor von Onsager is one of Hitler's best Aryans. If you are lying, it will not bode well for you. Do you understand me professor?"

Emil nodded yes, saving his life that day.

"Sergeant Schwartz, take this man back to his cell," Hans ordered. "And get him cleaned up and see to it that he is well fed. I will take him to Bonn. Make the arrangements as soon as possible. Also, if you know what is good for you, I would keep everything you heard today to yourselves. We need to corroborate our good professor's story. Have I made myself clear?"

"Yes sir!" both Sergeant Schwartz and Corporal Bailin barked out as they dragged the semi-conscious body of Professor Goldstein away.

***

Sure enough, to Hans' delight, there were Jews still living in Bonn who remembered Belinda's grandparents attending synagogue services before the fire at the registry. "The family changed their name and converted to Catholicism before leaving Bonn," one victim of Hans' interrogation confessed before being hanged. Hans even found an old photograph of

Lise's great-grandparents taken during their wedding ceremony tucked away in the basement of their old synagogue. Excellent, he thought to himself. He had everything he needed to start his game.

Within the week, Hans had arrest warrants drawn up for Dr. Reber, his wife Belinda and their two young daughters Olga and Nanerl. As soon as the papers were signed, he sent several of his most trusted SS thugs to round the Rebers up. If Hans knew Heinreich, though, he expected Heinreich to do something foolish—even perhaps to attempt to escape Germany with Lise. "Under no circumstances," he told his men, "are the Berlin Rebers to be rounded up until I have Heinreich and Lise in my hands at Augsburg. I don't want a frantic phone call from Berlin to alert my two lovebirds."

As for Heinreich and Lise, Hans arranged to be driven to Augsburg with a few of his closest SS officers. He wanted to see the look in Heinreich's and Lise's eyes as they were torn apart. Lise would be shipped off to Dachau to join her family, while, best of all, Heinreich, in order to keep the Rebers alive, would be forced to carry on with his military duties—check mate.

Hans and his four SS thugs arrived in Augsburg just before sunset. The base commander—a lank, middle-aged oberst with graying hair—was shocked that SS men had come after Heinreich and his wife, the most ideal, happy and delightful of couples. "Oberst Eichmann, we have come to arrest Hauptman von Onsager and his wife," were the first words out of Hans' twisted mouth when he walked into the oberst's office.

The oberst didn't even raise his eyes from the document he was reading. "Nonsense," he declared. "Show me the arrest papers. There must be some mistake." Unfortunately there was no mistake in Hans' arrest papers. They were official.

"Under what charges are you arresting my best officer?" the oberst asked.

"That is not for you to know oberst," Hans replied.

A few minutes later, Oberst Eichmann had himself, Hans and his four SS thugs driven to Heinreich's home, a smallish, but tidy cottage located a few kilometers away from the secret installation on the outskirts of Augsburg.

Hans and his men spilled out of the oberst's staff car with their weapons drawn. Two of the SS storm troopers kicked down the doors of the von Onsager household and ran in yelling for everyone to drop to the floor in what appeared to be an empty house. A storm trooper found Heinreich sitting in his study. "Stand up you dog!" the storm trooper yelled at Heinreich, but Heinreich, wearing his Luftwaffe dress uniform, didn't budge.

As Hans walked past Lise's little flower garden and into the house, the first thing he noticed was a framed photograph of Lise, Heinreich and himself standing in front of the Castle Academy the day of their graduation. Hans picked it up and walked it to Heinreich's office.

"Stand up before Major Fritz!" another storm trooper yelled as Hans walked in.

"Let the Hauptman sit," Hans stated nonchalantly. "Look at us," Hans continued while showing Heinreich the photograph. "We were something then weren't we? So loyal. The three musketeers."

Heinreich, never removing his fuming gaze from Hans, didn't utter a syllable, even as his own commanding officer, Oberst Eichmann walked in.

"But look at us now," Hans resumed. Then he tossed the framed photograph against a wall, sending shards of glass flying all over the floor. "We are all broken up."

Heinreich, glaring with rage, didn't flinch.

"I would like you to know dear Baron von Onsager,"

Hans resumed now speaking with an angrier, higher-pitched voice, "that we have good things in store for Dr. Reber and his vermin wife and children."

Without removing his eyes from Heinreich, Hans picked up the telephone sitting atop Heinreich's desk and had the operator connect him with SS headquarters in Berlin. After about a minute, the connection to the requested number was made. Hans spoke but two words. "Get them," he said before hanging up. "Your in-laws," he then said to Heinreich, "will shortly be on their way to Dachau courtesy of the SS." Hans then pushed his face into Heinreich's face. "Dear Hauptman von Onsager, do tell us where your wife is. Did you not know you have been whoring with a filthy Jewess?"

Heinreich, with his blood afire, bolted up with a look of murder in his eyes, making Hans recoil. Heinreich would allow no one to speak of Lise that way, and before the storm troopers could stop him, he had punched Hans square on the face, busting Hans' lips apart and likely breaking the coward's nose. While Hans collapsed into visions of stars, Heinreich snatched out his sidearm, but the leading storm trooper stopped him before he could shoot Hans dead. Instead, as Heinreich was being pinned to the wall, his shot flew well over Hans' head. Oberst Eichmann, who hated the SS as much as anyone in the regular armed forces, wished the bullet hadn't missed the little weasel's head.

Hans, with the help of one of his thugs, straightened himself up and wiped the blood from his mouth. After regaining his composure, he withdrew his own pistol and, with Heinreich still pinned to the wall, pressed it hard against Heinreich's forehead. "Tell me Hauptman von Onsager, where is that Jew rat wife of yours?"

Heinreich, peering straight in Hans' eyes, let a defiant

smile creep across his lips. "She is gone Major Fritz. Mein father learned about your ploy this morning. He explained everything to me, including the shame that would be brought to the Führer himself for awarding Lise, a Jew, the award you could never win you filthy swine. As for my Lise, she is in France by now, far away from your filthy claws."

Peter Wohlthat, returned from Spain months before, had, after Heinreich's desperate call for help earlier that day, secretly flown Lise out to Paris in a private plane. Peter had pulled out all the cash he had on the spot and given it to Lise.

When Hans realized that Lise had escaped him, and his masterful plan had been foiled, he spat bitterly on Heinreich's face then tried to slap him, but Heinreich, with his seething eyes still fixed on Hans, caught Hans' hand in mid flight. "We shall see who has the last word in this matter rich boy," Hans declared. "For now, you will officially renounce your marriage to Lise."

***

Lise had an aunt and uncle on her mother's side living in Paris. Her Aunt Lillian (Lilo) had nearly been crippled thanks to a bout of polio during her childhood years. Her Uncle Elster, who had been a widower before meeting Lilo, was running on in years, but still loved to ski and hike the Alps.

When Lise reached their house on the outskirts of Paris— greatly surprising Lilo and Elster—she was still in tears, but she didn't forget to thank Peter for risking everything to steal her away. "I love you Peter," she said to him. "Please, please take care of my Heinreich."

Peter kissed Lise's hand. "I will baroness," he promised her before returning to the airport. He had no idea to what he would be returning back in Germany, perhaps SS storm troopers waiting to arrest him, or even shoot him on the spot. One only lives once, he thought to himself as he climbed into the air. Lise, meanwhile, waited for the arrival of her family.

The reunion was not to be. The car Minister Boltzmann had sent to the Reber household got snagged in a traffic jam caused by an auto accident. By the time the driver arrived, it was too late. The Reber females had been caught. Max, incommunicado, secretly treating patients in a Jewish quarter, wasn't at home when the doors came crashing down. As he entered his driveway later that day, he was snatched out of his car and taken to SS headquarters.

Within the hour Max was thrown into the cell holding Olga and Belinda. Belinda, he noticed, was in shock and had blood and gore spread all over her dress. Baby Nanerl was missing. "Where is Nani? Where is she?" he asked Belinda while he tore at her dress to look for the source of all the blood—a bullet hole. Belinda, no matter how many times Max shook her, didn't utter a word. She only stared blankly into space.

It was Olga, now twelve years old, who answered her father's pleading question. "They shot her daddy. When the SS men came to the house, crashing through the doors, she started crying and wouldn't stop. So they shot her in the head. I think the bullet is lodged in mother's shoulder. You should take a look at it father."

Belinda had been carrying Nanerl when an SS storm trooper shot her brains out. "Quiet baby, quiet please," she had pleaded with little daughter who was terrified with all the shouting that was going on about her.

"Quiet the little animal or I will shoot her dead!" one SS man kept yelling until he finally put his pistol to the little girl's head.

Nanerl, burying her head into Belinda's shoulder, screamed out in sheer panicked horror when she saw the pistol being pointed towards her—she had seen a Jew shot dead in the streets a few weeks earlier and had suffered horrible nightmares since that day. A loud report followed a second later, and Olga saw her sister slump into her mother's shoulder as if she were asleep. While urine seeped out of Nanerl's dead body, blood poured out of the bullet hole in Belinda's shoulder. "See," Belinda had said, "she is quiet now."

"After they shot Nani," Olga continued, "they took her from us, but I don't know to where daddy."

Now Max, understanding the dried blood, bone fragments and brain tissue caked on Belinda's body and dress, nearly passed out. All the futures he had imagined for his youngest daughter, her birthdays, her wedding, her children, flashed before his eyes, and then Nanerl was gone—stolen from existence by an animal with a gun filled with religious conviction that Jews were only animals suitable for extermination. Max collapsed sobbing into his Belinda's bosom, but after a moment of wailing, he wiped his eyes clear, finished undoing his wife's dress and tended to her wound. Loitering around the remaining Rebers, more then 100 Jews carried on with their caged lives in the holding cell.

***

Minister Boltzmann, realizing it would be best to get things out into the open, related the troubles of his stepson to his friend the Führer. At first, Hitler was furious with Wolfgang.

"Thanks to your ineptness, I award Germany's highest science award to a Jewess! I leave it to you extricate me from this embarrassment. For the moment, I want nothing of this leaked out. Do you understand Minister Boltzmann?"

"This, my Führer, is not completely up to me. There is SS Major Fritz to consider." Wolfgang reminded Hitler.

"I will order him to secrecy," Hitler replied, "but after this, it is between you and him to settle this matter discreetly. I cannot be trifled with such stupidities. We are at war you fool!"

Within the day, Minister Boltzmann offered Hans a proposal away from prying ears at a busy park near the SS headquarters compound. "Release the Rebers and I will see to it that you retroactively receive Doctor von Onsager's award. Do so and I will rewrite history, claming there was an error in the scoring of the panel. You will have what you want."

"Nein Minister Boltzmann, but I will meet you halfway. I will have the Reber women moved to Oswiecim in Poland as a sign of my good faith. When I am given the award that was stolen from me, then I will have them released into the custody of the Red Cross."

"Why Oswiecim? And what of Dr. Reber?" Wolfgang asked.

"Dr. Reber will remain behind as my insurance policy," Hans replied matter-of-factly. "As for Oswiecim, we good Germans are to prepare our first of several camps there for holding the Jews and other undesirables. We will use the camp to encourage other nations to accept these animals through the Red Cross." In 1940 the Germans would rename the city of Oswiecim to Auschwitz and build three camps there: Auschwitz I, Auschwitz II-Birkenau and Auschwitz III-Monowitz. Hans knew the true purpose of these places. They were to be extermination camps in what would become Hitler's so-called Final Solution.

"And how will I know you will honor your word Major Fritz?"

"Because I am an officer and a gentleman of the SS."

***

Lise didn't learn about the death of Nanerl until nearly a month after her escape, when she received a letter from her father sneaked out of Dachau by one of his old medical colleagues, Dr. Spalding, who then passed the letter on to Heinreich. Dr. Spalding was the very doctor who had saved Lise's life, and Heinreich thanked the aging man yet again.

After Heinreich read the contents of the letter, he wanted to burn it, but, together with his own letter, he passed it on instead to his mother so that she could have it forwarded to Lise via her old contacts in Paris. Heinreich knew that the news of Nanerl would devastate Lise, but he thought she had the right to know the horrible fate that had befallen her youngest sister at the hands of Hans' goons.

"Nani!" Lise cried out before swooning. Uncle Elster caught Lise in his arms while Aunt Lilo recovered the letter from the floor. "Oh no," she said when she read it.

If her Uncle Elster and Aunt Lilo hadn't stopped her, Lise would have returned to Germany as Heinreich had feared when he had the letter sent off. Instead, while Heinreich, Frieda and Wolfgang worked towards extracting the Reber's from Hans' control, the old couple persuaded her to work with the Red Cross and the French embassy towards getting Olga and Belinda the necessary refugee status.

***

While Minister Boltzmann began making the necessary arrangements to change history, Frieda managed to secretly pass significant amounts of cash to Lise through her extensive connections in France. Each week during this time, Hans would send Heinreich a menacing letter stating that if he and his mother the baroness tried anything really stupid, he would personally see to it that the remaining Rebers would go to the gallows. Heinreich, Frieda and Wolfgang, where severely pinned down.

Hans reveled in the power he held over Minister Boltzmann, the baroness, and most especially Heinreich. Heinreich, meanwhile, prayed each night that Lise understood the situation—that he and his family truly had their hands tied. He didn't want Lise to stop loving him, think of him as a cold blooded Nazi, and ultimately hate him. The thought mortified him.

As for all the letters Lise wrote on behalf of her family to the European embassies and consulates in Paris, nothing ever happened. "We are sorry," the return letters would begin, "but our country has already accepted our quota of Jew émigrés for the year. Since your father, as you state in your letter, is a well respected medical doctor, we feel confident that he and your family will find refugee status elsewhere, perhaps in the United States. We wish you the best of luck."

Lise, crying bitter tears of despair, would crumple these letters with an angry face and toss them into the garbage pail. How much longer can they survive in Dachau she would wonder to distraction. "Pray for them and keep trying," were the only two things Uncle Elster and Aunt Lilo could say to her.

"Pray and have faith," were only things Belinda could think to say to Olga as the weeks in Dachau turned into months. "Pray that all of this nonsense, all of this hatred comes to an end sometime soon."

Young Olga couldn't find her faith however. Every day she saw brutal atrocities committed before her very eyes, hangings, beatings, shootings, strangulations, and so on. Every day someone would go mad and try to climb over the barbed wire fencing only to be shot by a guard in a watch tower. Every day the stench of rotting death filled her little lungs.

At night, while her father tried to care for the victims of the SS as best he could in the cover of darkness, Olga did her best to help her mother. "Don't cry anymore mommy. Nani is sleeping with God," she would say to Belinda while stroking her mother's hair. "And Lise and her husband are outside of this and doing their best to get us out of here." But Olga didn't believe her own words. Much less did the twelve year old girl believe in God. There were only monstrous devils in the world.

The year 1939 thus ended without any success for extracting the Rebers from Dachau. Hans' stranglehold on them was too tight. On New Year's Eve, Heinreich received a telegram from Hans. *"If your step father doesn't hurry up with my award, I will soon have the Rebers hung, most slowly, most painfully, one by one before your eyes."*

Heinreich, crumpling the letter, swore to himself that he would kill the little bastard one day. He knew that for now, at least, the war between himself and Hans was at a stalemate. His heart sank when he realized that he would likely not be able to see his beloved wife until after the war ended in a year or so.

***

As 1940 progressed, communication between Lise and Heinreich grew increasingly sketchy. By April of 1940 the Nazis had invaded Denmark and Norway, and by 16 May, the German Wehrmacht broke thru the French defensive lines on the River Meuse at Sedan. By 13 June, Paris had fallen to the invading German forces, and refugees by the tens of thousands poured onto congested roads in cars, on carts pulled by horses, by bicycle, and even by foot. It was a maelstrom of human chaos.

After the fall of Paris, all of Lise's and Heinreich's established lines of communication ceased to exist. The two lovers could only look up to the stars at night and wonder if they were sharing the same site.

During these weeks of turmoil in France, Lise, her Uncle Elster and Aunt Lilo decided to flee to Yugoslavia by way of Italy using Frieda's money, false passports they had purchased from a forger, and an expensive car borrowed from a French count with whom Frieda had once had an affair. The drive to Italy proved to be relatively uneventful despite Aunt Lilo's presentiment that there would be trouble with the Italian border guards. "They will stop us," she kept repeating. "Even if they do, our papers are in order. Now shush with this!" her exasperated husband would tell her. Lise, sitting in the back, feeling lost, didn't know what to think.

Once the trio entered Italy—having gotten over the border without incident—Lise, her uncle and aunt dropped speaking French and started passing themselves off as affluent German tourists. The strategy, as Germans were certainly welcome in Mussolini's Italy, especially ones driving in an expensive car and carrying lots of German cash, worked.

Lise, using her wits, had her uncle drive to many of the exclusive places that she and Heinreich had lodged at during their tours of Italy—Italy had been Heinreich's favorite country to vacation in. Most everyone at these more exclusive palazzos recognized the pretty young lady at first sight and asked her how her husband the baron was doing. After showing them her wedding ring with a smile, she would tell the clerks that her husband was doing quiet well, but that, being an officer in the Luftwaffe, he was occupied with his duties.

It was at Milan's main rail station where Lise's and her uncle's and aunt's luck ran out. Months before, in anticipation of the invasion of France, Frieda and Heinreich—who had figured out how to get smallish packages past Hans' SS men—had arranged for a package of money to be left at the station's safe thanks to the help of Frieda's long time friend, the Contessa Fredericka. While Lilo and Elster waited in their car, Lise made her way to the safe.

No one at the station—not even the station police—was any the wiser that Hitler, needing to meet with Mussolini, planned to arrive at the very same station unannounced until a mere half hour before his scheduled arrival. It was a matter of security to do so. Hitler's people simply didn't want communist terrorists to have enough advanced notice to plant their bombs.

Lise had only just finished recovering Frieda's package from the safe when the local police, who had gotten word of Hitler's impending arrival only a few minutes beforehand, began to arrive at the station in droves. Within minutes scores of police cars piled into the area, and policemen by the dozen began to set up roadblocks.

As Lise walked the long length of the station towards where her aunt and uncle had parked, piles of local SS men began to fill the station the minute Hitler's train stopped. The sight of these armed goons frightened her half to death. For a moment, she thought the SS men had come for her, but then realized this was ridiculous.

"What is going on? Why are there SS men here?" Uncle Elster asked Lise when she reached the car.

"I don't know uncle, but we must leave quickly. I have the money."

Before Uncle Elster could turn the car around and head for the exit, the twenty or so cars that were ahead of them were stopped. A dozen Italian policemen and several heavily armed SS security officers began walking down car by car and asking for people's papers.

"Mein God, they are going to shoot us," Aunt Lilo uttered.

"Calm yourself please Lilo," Uncle Elster admonished her. "We will be fine if you just manage to relax."

All Lise could say as she watched police and SS men ask for papers was, "I hope we get an Italian policeman and not one of those SS pigs."

A tense ten minutes after the first inspections began, an SS leutnant walked up to the car ahead of the one Lise, Elster and Lilo were in. He asked the family inside—a husband, wife and two children—for their papers. When the papers were not to his satisfaction, the SS leutnant withdrew his sidearm and asked the family to step outside of their vehicle. There was a look of dread and mortification written on their faces as they stepped out. One of them, a frightened little girl about Nanerl's age, looked Lise straight in the face before they were all escorted off into a truck.

Lise's wish for an Italian policeman was granted when a portly Italian corporal walked up to their car. "Le carte," the sweat soaked corporal asked in a hoarse voice. Uncle Elster complied, quickly handing all three passports to the corporal. A few seconds later the corporal returned Uncle Elster's and Aunt Lilo's papers, but there was something he didn't quite like about Lise's papers, which he kept staring at.

Lise, who spoke a decent Italian, asked the sweating

corporal what the problem was. He then asked her why she had a different name than her parents. Smiling, she explained to him with slow words that Uncle Elster and Aunt Lilo were not her parents, but were her aunt and uncle. The corporal either didn't seem to accept or understand this. Instead, he beckoned a nervous looking private armed with a rifle to come and stand guard over the car while he took Lise's papers to his commanding officer.

The tension in the car became nearly unbearable as the minutes passed by at gunpoint. Finally an Italian captain and an SS leutnant walked over to their car with the sweat soaked corporal following behind them. The SS leutnant was the very same one who had asked the family in the car ahead to step out, and he looked upset.

Before any of the officers could utter a word, Lise preempted them. In a perfect, haughty German she addressed the SS officer. "Thank God! I'm so glad to see a German face in this country full of idiot monkeys. How could they possibly be our allies?" Lise began.

The SS leutnant smiled. "What is the problem here?" he asked Lise.

"I am the Baroness von Onsager, as my papers clearly state leutnant. I am newly wed to Baron von Onsager, and I am on vacation with my aunt and uncle while my husband, a huaptman in the Luftwaffe bombs the French." Lise passed the leutnant a photograph of herself being kissed by a happy Heinreich in front of a German fighter at the Augsburg flight line. "The idiot corporal," she continued, "does not seem capable of understanding why my name should be different than that of Uncle Elster and Aunt Lilo Reber."

The SS officer turned his gaze to Uncle Elster. "And what brings the family to Italy?"

Uncle Elster, picking up on Lise's line, replied, "You heard my niece. Her husband is busying bombing France. Our chateau unfortunately is too near the fighting for our comfort. We fully expect that once you and the Wehrmacht have tamed that country, we should be able to return to our—I expect—unscathed home."

The SS officer had heard enough. "My apologies to you and your wife, and most of all to you Baroness von Onsager." The leutnant then clicked his boots and uttered, "Have a nice day and Heil Hitler." After he returned Lise's papers, he saw to it that barricade was opened for them. A few seconds later the portly corporal, looking cheated, raised the barricade and allowed the Rebers to pass through.

"I'm glad," Lise stated as they turned away from the rail station, "that we kept our original names on those false passports. A lie—as Hitler himself tells us—is best perpetuated when it is mixed with truth."

Uncle Elster, who burst out laughing added, "and when one doesn't have to strain to remember one's haughty title, dear Baroness von Onsager."

Later that day on the outskirts of town, Lise, her uncle and her aunt stopped at a prearranged café to meet with the Contessa Fredericka, one of Frieda's dearest friends from prep school. Lise—who hadn't seen the lady since her wedding in Capri—was both happy and sad to learn that negotiations were still going on between Minister Boltzmann and Hans regarding the release of Olga and Belinda. Her loved ones were alive, but still in the hands of the monster Hans.

"I'm so sorry for you my poor dear," the contessa told Lise while wiping a tear.

"Hans is toying with my Heinreich and getting back at me out of envy," Lise replied. She knew that if Hans ever lost control over Heinreich, it would likely cost him his life. That's why, even if the bastard ever released Olga and Belinda, he would keep her father behind.

At the end of their meeting, Lise gave the contessa a brief note to pass to her Heinreich. Lise then told the contessa their general plans to get to Yugoslavia. "If we succeed, we will send you notice. Please get word to Frieda of thanks on my behalf, and tell her how much I...love her son." Lise choked on these words, but managed to thank the teary-eyed contessa for her help.

"I promise," the contessa said. "And I will pray to God for you as I look forward to hearing from you soon."

***

The further Lise drifted away from her native Germany, the more she began to wonder if she would ever see her loved ones again. The lines of communication that linked her back to what had been her home world were disintegrating. From days, letters and notes began to take weeks to reach Germany. Soon, she feared, her letters and notes would begin to take months—as if she were back traipsing around the New World of the fifteenth century—to travel to their destination.

The last letter Heinreich received from Lise arrived in late July through Frieda's collapsing network. The letter had been passed to the Contessa Fredericka, who forwarded the letter to her nephew Alfonso in Rome, who then got the note to an old college friend living in Alsace, France, who then, after his two year old boy nearly destroyed it, finally passed it on to Frieda in Berlin. When it got to Heinreich's hands, the letter was crumpled, soiled, and looked like it had gone thru hell—Lise's hell—to get to him.

Alas, I fear the worst as we enter deeper into Yugoslavia. I fear

I will never see mother or father again, nor poor Olga, and that your and my moment in the sun is at an end.

Thank Wolfgang for his very brave efforts to free mother and Olga. Thank your mother for the package at Milan. We should have no need for cash thanks to her generosity. Do not blame yourself my dearest. I know that Hans is the one who holds my family's fate in his hands. I understand that even if mother and Olga are released, with my father trapped in Dachau, you will feel compelled to remain behind, pinned down by the monster. Please don't do anything rash. Keep yourself alive.

We will do our best to settle somewhere safe. I will try to send word from there if Contessa Fredericka is still around, and still able to help. With all my life, your dearest Lise. PS - Promise me to burn this letter as I fear Hans will find it. Remember me always with happiness as I remember you every moment of my life. I love you dearest husband. You are my life.

Heinreich, in tears, folded the little note and tucked it into his billfold. "I will kill you Hans," he muttered to himself.

***

Between 10 July and 31 October of 1940, the Battle of Britain raged on. Peter Wohlthat suffered a serious injury fighting over London, but managed to fly his damaged plane back to Germany, where he was promptly grounded. Heinreich, now promoted to major, saw to it that Peter was transferred to his secret test pilot squadron in Augsburg. Heinreich was in desperate need of a friend.

When Peter arrived at Augsburg two months later, still wearing his leg cast and arm brace, Heinreich had him promoted to hauptman. Peter appreciated the gesture, but he told Heinreich that he wanted to return to the front lines as soon as possible. "My place is in the skies shooting down our enemies." Heinreich refused his friend's requests for combat duty each time.

No matter how many times Heinreich tried to get assigned to the front lines, he was himself kept back by his own superior officers. His work in the Augsburg tunnels was too important to Germany he was told. He was, after all, helping Messerschmitt design and perfect Germany's first jet fighter, and such a technology was poised to wreak havoc on allied air forces.

"I promise you Peter, the moment I can get out of this damned place, I will take you with me straight to the front lines and straight to hell," Heinreich often told his friend at the officer's club.

"I know you will," Peter would reply. "Have you heard anything from Lise?" he would then duly ask.

Heinreich could only shrug his shoulders each time he heard this question before giving his usual reply. "No. Not since many weeks now."

"Don't worry. She is fine," Peter would then reply.

"Perhaps Peter, she hates me. I can't stop hating myself for being unable to help her family. Hans has such a strangle hold on them, and I helped make the little bastard who he is."

Peter would then rest his hand on Heinreich's back. "Somehow," he would tell his friend, "I have a feeling things will turn out well. I can't say when, but sometime from now. Have strength and faith my friend. You know that Lise is a smart, resourceful girl."

After Lise got past the Italian border and into Yugoslavia, Heinreich had no idea as to where exactly Lise was, or even, since receiving her last letter, if she still lived. It was as if Lise and her relatives had dropped off the face of the Earth.

Heinreich's spirits worsened when French and Italian

Jews started pouring into Germany by the tens of thousands. With the help of France's Vichy government and sympathetic Italians all too eager to push the Jews out of their countries and steal the spoils, an endless supply of cattle cars were stuffed with Jews and shipped off to central depots, and from there onto concentration and forced labor camps.

Was Lise among them, Heinreich would ask himself each time he saw a train pulling a long string of cattle cars filled with the damned. The thought gave him nightmares. When he sent a letter expressing this fear to Frieda, Hans' SS men intercepted it. Not long afterwards, Hans started dropping Heinreich a note on a weekly basis. *"Dear Heinreich, Lise is not in this week's batch of cattle cars. I'll be sure to let you know as soon as I find her old friend."* Every time Heinreich received one of those vile notes, he promised himself that he would strangle the little bastard. "And I will hold him for you while his eyes pop out of their beady little sockets," Peter would be sure to add.

***

Lise, Uncle Elster and Aunt Lilo made it into Yugoslavia, but not much further. Their luxury car broke down on some unpaved country road that was not on any map. Before them, behind them, and all around the lost trio stretched rolling plains of grasslands. With Lilo essentially wheelchair bound, the three of them, tired, thirsty and hungry, were at a loss for what to do.

"We are going to die here," Aunt Lilo concluded. "One day someone will find our automobile and our bones beside them."

Lise for some reason—though she was feeling scared—found this picturesque idea somewhat preposterous. She burst out laughing while Uncle Elster and Aunt Lilo looked on thinking the poor girl had gone mad. "I suggest," she started to say, "that we write our names using a pile of stones for the benefit of Aunt Lilo's somebody who, someday, will find our bones." Uncle Elster tried to reach out to Lise, but she pushed his hand away. "I'll tell you what we are going to do. We are going to live for as long as we can until there is no more dignity in bothering to suck in another breath. You Uncle Elster, and I, will push Aunt Lilo's wheelchair on this road until we either reach the end of the world or someone's house. Am I understood?"

Both Uncle Elster and Aunt Lilo nodded their heads in emphatic agreement.

"Good," Lise added. "Then let's get started. I'll push first. Uncle Elster you carry the valise with our absolute essentials."

The trio pushed on without stopping until the sun began to set before their faces.

"Stay with Aunt Lilo," Lise told her uncle. "I'm going to put to use some of the things I learned from the Nazi survival teachers at the Castle Academy and find us something to eat."

"Wait dear," Aunt Lilo interrupted. "I think I see someone coming up the road."

A man sitting atop what appeared to be a mule drawn cart was indeed coming up behind them at a very slow pace. The man, who stopped his wagon beside the lost trio, turned out to be a kindly farmer, strong looking with graying hair and sunburned skin. He didn't speak a word of German or Italian. Using hand gestures, the farmer explained that he had passed by a broken car, searched for its owners, and finding none, had then pressed on.

Lise felt stupid.

"No dear," Aunt Lilo told her after taking Lise's hand. "You did the right thing."

The farmer then offered Lise and her elderly relatives a lift to his place, a sprawling farm that would not have looked out of place a thousand years into history. It was filled with scampering chickens, pigs, cows, dogs, cats, horses, goats and a little army of children from babies to young adolescents who came out to stare at the strange strangers.

The farmer's wife, seeing the tired faces of Lise, her uncle and her aunt, did her best to shoo the children and goats away as the sun began dipping beneath the horizon. The thin, but brawny woman then ushered them into her kitchen and offered them fresh water, while her own plump mother busied herself with the cooking. The smell of ham roasting in the oven made Lise's mouth water with anticipation. The dinner that she, Elster and Lilo shared with the farmer's family, was the best dinner she had ever had in her life.

The following day, the men folk retrieved the broken car with a pair of mules. They tried to repair it, but it was impossible to get the parts required by the luxury machine from anywhere but Germany. The end result was that Lise, Lilo and Elster were left stranded.

"Perhaps this is for the best," Uncle Elster suggested standing beside the dead car. "It's not as if we were going anywhere in particular."

Lilo, exhausted and resigned couldn't disagree.

The farmer, his wife and his eldest sons knew that the young lady, her aunt and uncle, were German Jews. They also knew, if they were to be true to their faith in God, that they could not send the refugees packing. Many Yugoslavians in the cities—in anticipation of currying favor and collaborating with Hitler's troops, whom they fully expected to invade—were already turning against the Jews of their nation, making lists of their addresses and assets. The farmer and his wife offered

Lise and her uncle and aunt a place to stay. The trio accepted with great humility. They would stay in the barn, out of sight, until the men dug out a secret, hay covered chamber beneath the barn floor.

"I will enjoy sleeping in the barn," Lilo stated. "It reminds me of my schoolgirl years, before the polio, when I lived on my father's farm."

"I've been a city girl all my life," Lise confessed. "I suppose I will learn to milk cows and make butter. I will make myself as useful to our hosts as I can." That afternoon she did her first useful thing. She fixed the farmer's old radio.

As one week piled into another, the one thing that upset Lise most as she and her relatives settled into a life of hiding in their secret pit during the daylight hours, was the refusal of her hosts to forward her letters to Italy. The family was very afraid of getting caught harboring Jews. They had good reason to be afraid. Things for Yugoslavia—according to the Yugoslavian news broadcasts the farmer's family listened to on the old radio—were looking increasingly grim given the advances of the Axis powers into that part of the world.

Within a few months of their entry into Yugoslavia, things darkened precipitously when Romania joined the German, Italian and Japanese Axis, almost completely surrounding the country's borders with hostile powers. When Bulgaria signed on to the Axis powers some months later, Yugoslavia became completely surrounded, and every Yugoslavian knew that it would not be long before Hitler's tanks would roll into their country. During this period, Lise picked up the native language and began giving the farmer's children lessons in science and mathematics. She even built them an old telescope out of junk parts, and the children marveled at seeing the red storm of Jupiter and four of its brightly lit moons, another world beckoning her to it.

Often during these dark months, Lise marveled at the courage of the family, putting themselves at risk for her, her aunt and uncle. All over the European front, small numbers of good people were doing their part to save humanity—families and little girls like Anne Frank—in dank pits, in attics, in hollowed out walls.

Including the loss of Yugoslavia, the following year (1941) spelled more disasters and defeat for the British and their allies. The Germans, throwing the lessons of history to the winds, invaded the Soviet Union in June. By the end of the year, Japan attacked the United States at Pearl Harbor, Hawaii, sinking many ships but the all important American carriers they had wished to destroy. President Roosevelt declared war on Japan the following day, and on 11 December 1941, Germany and Italy declared war against the United States. The sleeping giant—as Admiral Yamamoto feared—thus awoke. Japan had struck a hornet's nest of industrial might and the ire of a peoples who didn't much like being messed with.

**

The Third Reich's Educational Minister signed the orders with bitter sweetness. Wolfgang's people were instructed to destroy all records of Lise's existence at the Castle Academy, her dissertation work at university, and her having won the first Hitler Science Award.

"With the stroke of a pen I've made it so that our daughter-in-law has never existed in Germany's educational system," he penned his wife with a heavy heart. "With another stroke of a pen, I've rewritten history, yielding to our petulant Hans his much coveted award. This whole affair has made me sick to my stomach. Now let us see if he does as he has promised to release the Rebers."

Wolfgang passed a facsimile of his orders to Heinreich, who, even though he realized this was the first step in saving Max, Belinda and Olga, nevertheless felt his blood boil. "Thank you father," Heinreich told Wolfgang over the phone.

When Hans received notice of the orders issued by Minister Boltzmann, he was tickled pink. At last I've won my well deserved prize, he thought to himself as he reread Wolfgang's orders behind his enormous desk.

That same day, Hans began procedures regarding the release of the Reber survivors. He smiled when he, in turn, signed his orders. "There's a difference isn't there dear Eva," he asked his latest personal assistant, "between procedures and release?"

Leutnant Eva Reinhardt, not quite sure what her boss meant, smiled. "Indeed sir, there are." Eva was an attractive blond with high cheek bones and piercing blue eyes who spoke with a cheery, but throaty voice.

"As soon as the Auschwitz camp is ready dear Eva, have the Reber women relocated as soon as possible. Also, send a letter to Educational Minister Boltzmann that he need not worry about the stain of his stepson having married into a Jew family. I have seen to it that the marriage records have, shall we say, been lost. Moreover, tell the minister that, once they arrive at Auschwitz, I will soon make arrangements with the Red Cross to transfer the Reber women to the English. Oh yes—before I forget—please make arrangements for dinner."

"With whom sir?"

"With whom do you think Eva?" Hans asked with a wide grin. Eva smiled. "Yes sir, at your favorite place sir." Hans did not actually care for Eva, or find her particularly attractive despite the size of her bosoms. He considered her to be dim witted, but he knew that other men slobbered after her, SS men in particular, and it gave him—still boyish looking—a sense of power over them. Whenever he didn't have a business dinner, he was sure to take Eva to the finest restaurants where many ranking officials of Hitler's staff would also dine. He knew full well that Eva got him noticed. I am a genius, he liked to say to himself during these days of triumph.

The day Belinda and Olga were torn from Max's life, boarded like animals onto a crowded cattle car on Hans' orders, it rained. "Have high spirits dear wife. Soon you and Olga will be in England," Max told his wife at the edge of the cattle car. He hoped it was true that his wife and Olga were to be released to the Red Cross, and didn't dare let on that he reserved some doubts concerning Hans' actions.

"If we don't see you again on this Earth," Belinda told her husband as she was forced to let go of his hands by an SS guard, "always remember that Lise, Olga and I, and Nani in heaven above us, love you. I could not have asked for a better man to give children to. I love you Max, my dear husband."

Max, shedding tears of anger mixed with anguish, threw his arms around Belinda and his daughter at the base of the ramp. He kissed them both one last time before another SS guard, using his rifle, pushed him away so that he fell into a muddy puddle. "I love you both! I love you both so much!" Max yelled at them as the doors of the cattle car were opened and Belinda and Olga were forced inside along with a hundred other miserable people.

"I love you daddy," were the last words Max heard Olga speak as he stood up, just before the cattle car doors were shut closed and the train steamed off for Poland beneath gray skies. She looked so small and vulnerable to her father as he tried to burn her image in his mind.

***

"Take heart my son," Minister Boltzmann wrote Heinreich when he got official notice of Belinda's and Olga's departure for Auschwitz. "They are gone to Poland, and soon this mess will be over for them."

It was a happy day for Heinreich when he received his stepfather's message, but he despaired more than ever that he still could not get hold of Lise. "Is she dead?" he started to wonder. "No," Peter would assure him. "You will see that she is fine hiding somewhere in Yugoslavia as the contessa suspects." Keeping his faith, however, was proving itself to be increasingly difficult for Heinreich as the months seeped by.

For many weeks Heinreich and Frieda tried to pass Lise the information about the Red Cross exchange in Poland via their Italian contacts. Their efforts were for naught. It was frustrating, but Heinreich both understood and appreciated the risks these people took on his behalf. Every act of defiance where Nazis and their cronies had power—even just passing a note into or out of a prison camp—came with great personal risk for self and family. The SS, the Gestapo and their all too willing informants, skulked about everywhere.

"I wish I knew if Lise were alive, even dead mother," Heinreich would tell Frieda during these times. "The world is a mess at the moment my son. All we can do is hope and avoid giving in to our despair."

Lise, living in the Yugoslavian family's underground chamber with her aunt and uncle, had little notion as to what was going on in the world except through what the farmer's family told her. What news the large family learned about the world, they would learn only on Sundays, when they would attend church services in the nearby village. Once in a while, borrowing the farmer's radio, on certain nights when the weather was just right, she could tune in distant, static filled German, Italian or even English transmissions from the front.

"I think uncle that I will go mad here," Lise would tell Uncle Elster after her nightmares began. Lise had a terrible presentiment that her husband would get shot down and die, as his father Karl had, in a ball of fire.

"Lise dear," her uncle would tell her trying to wake her, "Heinreich is a strong, smart boy. He will make it through this war, and the two of you will thrive. You are both still so young, and this war cannot last forever."

Lise, in a cold sweat, slowly realizing where she was, in a dark pit dug into a barn, would smile. "Yes uncle. You are right."

***

The Red Cross, counter to what Hans had written to Heinreich, could do nothing to liberate Belinda and her daughter. By Hans' orders, the two women were to be held until the proper papers could be signed to his liking, and not one minute sooner.

A year of diplomatic impasse, with Hans holding all the cards, came and went at Auschwitz. One day, thanks to a clerical error that reached Hans too late, Belinda and Olga, who was beginning to bloom into a beautiful young woman, along with a hundred other girls and women, were marched to a "disinfection" room and given death.

When the group of one-hundred girls and women arrived at the "shower" chambers, they were told to strip off their clothes and pile them into categories. Everyone at Auschwitz had heard the rumors about the gas chambers and crematoria by then. A few of the women still held out hope as they undressed. Other women tried to do their best to help. "Disinfection is good," one older lady blurted out in a cheerful voice. "I cannot stand the lice anymore. They are driving me mad." It was her way of calming some of the younger girls who were near panic.

"Mother, do you believe that we will be sent off to the new camp after we are cleansed?" Olga asked her mother. This was the Nazi line fed to the would-be victims to prevent rioting and resistance.

"Yes darling. Of course," was all Belinda could say as the SS men began shoving the girls and women en masse into the "shower" chamber.

"Don't worry," Olga heard a green-eyed, blonde-haired SS woman tell her with a pretty smile. "It is only a shower."

The cold doors closed behind Olga's nude body, pressing her against others. There were so many bodies pressed together in the chamber that she found it difficult to breath. Belinda did her best to make space for her daughter as several minutes ticked by.

Olga heard the sound of a small engine starting. Panic struck the girls and women as gas, not water spray, started filling the chamber. Olga knew this was the end. She threw her arms around her mother and hung on as tightly as she could. Whatever was going to happen, she didn't want to be alone. All Belinda could think as more gas poured in—which smelled to her like putrid cheese—was that she didn't want Olga to suffer. "Be brave my little one," she told her daughter as more screams of panic burst out. "We go to meet God and Nani today."

The screaming, the clawing and the gasping in the overfilled shower chamber lasted about ten minutes, growing weaker by the minute, until all went silent. A few minutes later Polish workers—who had been waiting outside—entered the death chamber and began to remove bodies.

Olga, still clasped to Belinda, was forcibly pried from her

mother. Her face was contorted with agony. Her empty eyes stared into space as her body was piled on a cart atop other lifeless figures. Then she, together with a dozen other corpses, was rolled into an adjacent room where her hair was clipped to be made into expensive wigs. Belinda's teeth, the ones that had gold fillings, were extracted. Her hair was also cut off. An hour later, her and her daughter's bodies were incinerated into ashes.

Belinda and Olga ceased to exist while Max, back in Dachau, working as a camp physician, clung to his hope that his girls would soon be released to the English. He would have killed himself had he learned their fate, but Hans, protecting his means of power over Heinreich, his father the minister and his mother the baroness, continued with the charade, telling them that it was only a matter of signing the proper paperwork before he would release Olga and Belinda to the English Red Cross. Hans also kept up his charade to protect his own life. He had little doubt that Heinreich, if he should ever learn the truth, would try to kill him.

Two days after Olga and Belinda died in each other's arms, the war landed on Heinreich's lap. An allied bomber raid killed Frieda. The baroness and her driver had been headed to a bomb shelter when a stick of bombs began following her car down the avenue. The detonations were ear shattering as the driver tried to accelerate down the street. When the penultimate bomb exploded ten feet behind Frieda's car, it threw the whole vehicle into the air. It struck the ground and bounced once just as the last bomb detonated atop her car. There was little to recover of the wreckage and carnage bigger than a few small shards of blood-stained sheet metal.

***

Heinreich was in his office when the telegram arrived from his step father. "Dear son, your mother died yesterday. Her car was destroyed when an American bomb struck the vehicle. I am on my way to Berlin to handle things. We need to be strong for one another."

Heinreich let the telegram slip out of his hands. His blood boiled even as the first pangs of loss washed over his body. The Allies had been bombing Berlin since September of the previous year, and Heinreich had warned his mother to leave the city, but Frieda had refused to leave it when its people needed her most. She poured money and resources into the city's hospitals and shelters, and worked indefatigably to help the victims of the incessant bomb raids. "For this mein mother was killed?" Heinreich yelled turning red with anguish. "An American pilot killed mein father, and an American bomber crew has now killed mein mother, and here I am, a damned test pilot who has not done one damned thing to defend my country!"

The day Heinreich and Minister Boltzmann buried Frieda's few remains was the saddest day of Heinreich's life. It felt strange to him that all around him the people of the old city, walking around the bomb craters and the collapsed buildings, carried on with the normal exigencies of their daily lives despite the war going on around them. The women went out shopping for bread, the children played, and the old men read the newspapers. All Heinreich wanted to do was to close his tired eyes, fall asleep for a few years and wake from his nightmare at war's end. His mother was dead. His wife had disappeared into the jaws of a war that threatened to consume the whole of the so-called civilized world. There was no end in sight to the madness—only the bombs that kept raining from the skies.

Around the time Frieda was killed in the spring of 1941, Germany finally made good on its intention to invade Yugoslavia. The tanks, as had been long anticipated, rolled into the country and began wreaking havoc on the citizenry. Lise, Uncle Elster and Aunt Lilo made as if they had abandoned the farmhouse.

Before the German invasion, the trio would stroll about the grounds and generally try to be useful to their hosts. On a rare occasion, a neighboring farmer, or his wife or children would drop by to visit, exchange farm equipment, borrow a tool, and so forth, and these people were generally aware of the presence of Lise, Elster and Lilo.

After the German invasion, the surrounding neighbors were told the refugees had left, and Lise, Elster and Lilo only dared come out of their hole in the dead of night, lest some passerby might spot them. The loss of freedom and the fear of being caught by the Nazis now weighed heavily upon their hearts. It looked to them, from the little news they got, as if Germany and her allies would swallow the world whole. In the Pacific Ocean, Japan was pounding the Americans. In the Soviet Union, German tanks advanced on, inflicting crushing defeat after crushing defeat against the Soviet army. Greece fell a few weeks after Yugoslavia fell. In northern Africa, General Rommel's Africa Corps were decimating the British forces. Only Uncle Elster, who had lived in America for ten years, had faith the Germans would be defeated one day. "Americans are slow to anger, but provoke them, and you raise the ire of the most determined collection of mutts ever assembled in the world. I've seen it with my own two eyes dear Lise."

It was a bittersweet thought for Lise, knowing full well that her husband, if he hadn't already been killed, would have to fight for his country, and likely die doing so. She also pitied him. Heinreich was in the worst of moral dilemmas. If Heinreich betrayed his country, not only would he betray his country, but, thanks to Hans and his game of cat and mouse, his betrayal would cost the life of his father-in-law, if not also of Olga and Belinda. Lise couldn't bring herself to believe that Hans would ever release her sister and her mother.

***

When Heinreich heard the claxons warning of yet another Allied bombing raid flying high over his secret, underground base near Munich just a week after he had buried his mother, something in him snapped. He knew the Me 262 jet prototype that he was supposed to test later that day was fully armed and fueled inside its deep underground bunker facility. He put on his flight suit and headed down deeper into the main tunnels another hundred feet below the surface. When he reached the jet, he ordered the technicians to place it on the elevator that was used to lift the experimental jets onto the runway surface.

"What are you doing Heinreich?" Peter yelled out to him when he noticed what was going on.

"I'm going to fight," was Heinreich's reply with clenched fists.

A senior Luftwaffe oberst, who had been summoned by one of the technicians, showed up and accosted Heinreich. "What do you think you are doi—" the colonel tried to say, but Heinreich didn't let the colonel finish another word before he knocked the middle aged officer flat on his back unconscious.

"Are you mad?" Peter uttered when he reached Heinreich.

"Yes," Heinreich replied. "Are you here to stop me?"

Peter, squatting down to make sure the colonel was

okay, got up and looked at his friend. Heinreich was beet red with anger. "No. I'm going to join you. My 262 is fueled and armed as well." Heinreich didn't utter a word as Peter bullied the technicians to get the planes ready. "Hurry you dogs!" he barked at them. "The Allied planes will be flying overhead in a matter of minutes."

By the time the base commander knew what was going on, it was too late. Heinreich's and Peter's twin engine Me 262 jets were off the ground and climbing hard towards the contrails of the high-flying B-17 bombers. As usual, the cluster of bombers were flying in a tight formation at about 30,000 feet without a fighter escort. No Allied fighters had the range to fly with the bombers that year. When the bomber crews got sight of the two fighters approaching them from below, they couldn't believe how fast those things were moving. "They don't have propellers!" one gunner yelled out into the intercom with incredulity.

With Peter by his side, Heinreich picked out the lead bomber. At 500 meters he pulled the trigger and his four MK 30mm cannons started spitting out angry lead. Chunks of the bomber's nose were ripped off. Inside the lead B-17, the pilot and copilot were cut to bloody pieces while the gunners returned fired, but the gunners, not used to the speed of the propeller-less planes, missed wildly. Heinreich, gritting his teeth, full of rage, kept pouring fire into the lead bomber until it burst into bright orange flames.

Heinreich was so enraged he was going ram the bomber looming ever larger before his eyes.

"Heinreich, what the hell are you doing?" Peter asked over the radio, but Heinreich didn't reply.

A mere 50 meters before ramming into the bomber, Peter broke off as Heinreich pressed on at full throttle. All of a sudden, as one of Heinreich's 30mm shells hit one of the bomber's bombs, the smoldering B-17 exploded into a huge fireball of metal shards, smoke and light which Heinreich flew right thru. The decapitated head of the bomber's navigator bounced off Heinreich's canopy leaving a thick red stain of blood.

"You are insane," Peter called out.

Heinreich, continuing to ignore his friend, picked out another bomber. "Are you going to shoot down bombers, or are you just going to yell at me Peter?" Heinreich finally asked.

"Shoot down bombers of course," Peter replied.

All in all, after it was all said and done, Heinreich destroyed three B-17s, while Peter destroyed two more and severely damaged another one, sending a total of 45 allied aviators to their fiery, bullet-riddled deaths in the cold clouds.

Needless to say, there was hell to pay when Heinreich and Peter landed. The instant they stepped of their planes, they were immediately arrested by several military policemen and locked into their quarters to await their fate.

"I don't think we're going to have to wait long before we know how deep or how hot the water we are in is," Peter told his friend.

"I think you are right Peter," but I have no regrets. "We got five of the bastards."

Peter smiled. "Imagine that, getting into trouble for shooting down the enemy. Hell! I have no regrets either my friend."

Before the end of the day, Luftwaffe High Command sent then twenty-eight year old Oberst Galland to Augsburg to take names and roll heads.

"What the hell gave you two idiots the right to jeopardize your country's secrets?" where the first words out of Galland's mouth when he stepped into the small room where Peter and

Heinreich had been locked. "Do you know Göring himself wants to hang you two, literally? Speak up!"

If Peter was feeling remorseful, Heinreich wasn't. "When is fat Göring, whom mein father saved twice, going to use our secret weapons? After the war is over? We sliced through five bombers as if they were warm butter and we were hot knives."

Oberst Galland, who had 57 kills under his belt at the time, was of the very same opinion concerning "fat" Göring and the wasting of the Me 262. Of course, he didn't let Peter and Heinreich know that as he continued chewing on the two pilots.

Five more minutes of venomous vituperation ensued with Heinreich and Peter sweating bullets before Galland, his voice crackling, reached the end of his tirade. "If you two ever pull a stunt like that again, I will shoot you myself! Now, what do you two say about flying for me?"

Heinreich wasn't sure he heard Galland's question quite right. "As in flying combat?" Heinreich asked.

"What else? Flying mail perhaps?" Galland asked in a huff before walking out of the room.

From around the corner Peter and Heinreich heard the oberst yell as he walked away, "and I want full reports on how those jets did, and, after you make the arrangements, I want to fly one myself."

Breaking the rules turned out to be the best thing that happened for Peter's and Heinreich's careers. The following year Galland was promoted to general, the youngest general officer of the entire second world war. In turn, Galland promoted Peter to major and Heinreich to oberstleutnant (lieutenant colonel). He liked his two miserable "cowboy" pilots.

***

In 1942, a gas chamber was built in Dachau where Dr. Reber—who was still classified as a German sympathizer of Jews—did his best to help the sick and injured. More than a year had passed without hearing a direct word of his wife or daughters Olga or Lise. All his information came to him via Heinreich.

It depressed Max to no end to read from Heinreich's letters the news that the Allies were still suffering defeats all over the globe. The Axis powers seemed invincible, countering his childhood belief that good always triumphed over evil. Perhaps, he lamented, it just wasn't true.

To Max's great revulsion, a series of experiments were started on prisoners of Dachau later that year. Selected prisoners were infected with tuberculosis and malaria to test out several drugs. Others were frozen or put into a vacuum chamber to see what the limits of human tolerance were to extreme conditions. If the victims survived their experiments, Max would try to mend them.

Max's first glimmer of hope amid all the misery that surrounded him came when he heard from a new prisoner that the Americans had started a massive air campaign in Europe, and that Rommel had been pushed back by British and American forces in Africa. He was saddened, however, when he learned from Heinreich that Frieda had been killed in an American air raid. The baroness was an honorable woman who truly did her utmost for his family.

Lise, Uncle Elster and his wife Lilo learned of the American bombing campaign from their Yugoslavian hideout from a British radio broadcast. "I told you about those Americans," Uncle Elster proudly declared. "They are going to smash the Nazis!"

Lise, after milking the cows and cleaning out the horse

stables, started keeping herself busy by trying to work out a theory for the strong nuclear force, as well as tending to the needs of her crippled aunt. She tried to hope for the best because whenever she thought of her family—still likely in Hans' hands—she would fall into a black depression. The pang of not being with her husband, however, never left her.

When one day Lise realized from her work that it was possible to build an atomic bomb. She suddenly became frightened for the fate of humanity. "The Germans, if they figure this out first, will build super bombs," she told her uncle one evening.

Heinreich, flying combat missions against never ending waves of American bombers, living the life of an exhausted fighter pilot, never stopped thinking about Lise either. He kept a picture of her taped into the cockpit of his fighter, and didn't care what anyone might have thought of his having a picture of a known Jew on display. Only Peter managed to keep Heinreich somewhat grounded and sane.

Heinreich was promoted to oberst and Peter to oberstleutnant. The number of American bombers filling German skies only continued to rise that year.

"Do you think we can win this war?" Peter once asked Heinreich after a hard fought mission in which seven German fighter pilots had lost their lives in a battle against 500 American bombers.

"I don't think about such things," Heinreich had dryly replied.

As the war dragged on, the two young men witnessed a lot of carnage, both American and German, and began to realize the true cost of war. It became a job just to wake up for another day of fighting.

By this point in time Hans' career eclipsed that of his

father's. It seemed there was no limit to how high the "boy genius" might one day rise within the political ranks of the Nazi party, even as Germany began to show signs of cracking against the Allied onslaught.

Hans' career took a quantum leap forward when he married the youngest daughter of General Schmitt, another Eva. Ten months later Albert Einstein—who could not possibly know it—became a great-grandfather to little Axel Schmitt.

Hans forbade Himmer from ever seeing his grandson. "Do you wish to call him a rat Jew boy as well father?" Hans asked Himmer the day Himmer tried to see Axel. Constance loved Axel, and the fact that she was a grandmother.

When Minister Wolfgang suffered a stroke in late 1943, he became wheelchair bound and was forced to stay home with a nurse. He lost the use of half his body and could hardly speak a word. He shot his brains out Christmas Eve 1944, leaving a mess of family business for Heinreich to settle. Wolfgang knew that the war was lost, and that no woman would ever love him again.

***

On June 4 1944, after slugging their way up the Italian Peninsula, U.S. and British troops entered Rome to cheering crowds. Two days later, Allied forces landed on the beaches of Normandy on D-Day. By late August Paris was liberated, and by October, the Americans invaded the Philippines. That same month the Yugoslavian capital of Belgrade was liberated by Soviet forces on 20 October 1944.

After over three years of hiding in an underground chamber, Lise, her Uncle Elster and her Aunt Lilo, looking pale from a lack of sun, clambered out of their pit and gave themselves up as "French Jews" to a column of advancing

Soviet soldiers marching towards Germany. It was a grave mistake.

Looking thin, but still attractive with her hair now grown quite long, the first thing the soldiers tried to do was rape Lise. Fortunately, the officer in charge, a young captain who had been a teacher at a technical school, stopped the men before they could get very far. Unfortunately, Uncle Elster, who had tried to defend her honor, had had his head caved in with a rifle's butt for his efforts. Aunt Lilo, shocked at the site of her husband's split head, began to beg one of the soldiers to shoot her, but Lise, quickly grabbing onto her sobbing aunt, managed to calm the old lady's wailing down to a muted moan. "Quiet Aunt Lilo, or they will shoot us both."

Not quite knowing what to say to Lise or Lilo, the young captain, who spoke a decent enough French, and who was obviously feeling ashamed, had the old man's body taken away. Hours later, once the captain's commanding officer discovered that Lise and her aunt were actually German Jews fluent in German, the women were ushered off to division headquarters. The Soviet general staff was in need of good translators for translating the statements of German prisoners of war. Before Lise and Aunt Lilo were driven away to division headquarters, the captain, trying to make amends for the behavior of his men, gave the pair of women a thermos filled with hot coffee and some rye bread. Neither Lise nor Aunt Lilo got a chance to thank the family who had cared for them for so many years. The family understood. They buried Uncle Elster among their own.

Lise's and her aunt's newly found work at division headquarters was horrific. They were tasked to translate the awful, agonized howls and pleas for mercy of young German soldiers suffering the most egregious tortures. The Soviets wanted any useful morsel of information the German prisoners might provide them before they marched the captives outside and summarily shot them in the head.

Lise and Aunt Lilo would have no choice but to watch as the torturing was conducted. They tried not to look at the faces of the young men as they were dragged outside pleading and squealing for their lives. After the prisoners were shot, the lifeless corpses of these young men, who had been in the very flower of life, were then left to rot outside.

Things only grew worse as the Russian column Lise and Lilo were with advanced deeper into German territory. The more German prisoners were taken, the greater the brutality against them was perpetrated. Whenever the Soviets captured too many German soldiers at a given time, they would lock the soldiers into a church house, or some such wooden structure, and burn them alive as the Germans had done to the Russians the year before.

As 1944 progressed into 1945, Lise and her Aunt Lilo, who were now eating on a regular basis, began to look and feel healthier of body, if not in spirit as they were dragged along ever closer to their native Germany by the Soviet column. On New Year's day 1945, they were told the division was heading to Berlin to pay Hitler a little visit. "Once Germany is defeated," the two women were told that night, "you will be freed to return to your homes." This, along with a shot of vodka, was their Christmas gift.

Lise, who wanted nothing more in the world than to return home to her family in Berlin, and to find Heinreich if he still lived, accepted the spirits but didn't drink them. Neither did Aunt Lilo. They wanted nothing to do with the barbaric Soviets but to get away from them.

***

7 May 1945 was the day before Germany officially surrendered to the Allies. On that day, Hans, his wife Eva and his son Axel, along with dozens of other SS and Gestapo Nazis, were on a small passenger ship steaming towards Argentina after having gouged out their SS tattoos. General Schmitt, having committed suicide, was not among them.

At the same moment Hans was looking out at the open ocean, Heinreich and Peter were flying side by side in their Me 262 jet fighters at full thrust. "Peter, you box that American bastard my way," Heinreich yelled out over the radio. "Pin him against the bombers."

Heinreich and Peter were pursuing an American P-51 Mustang. That the second world war was winding down to its end didn't stop the friends from rushing to claim a last few Allied victims over the skies of Germany.

At 1,200 meters Heinreich squeezed off several rounds but missed the wily American as Peter tried to push the Mustang towards Heinreich's guns. As good a fighter as a P-51 was, it was still only a propeller-powered plane, and Heinreich had two powerful jet engines to power his fighter. No matter what the desperate American pilot did to get away, Heinreich would cut him off and close the distance to his target.

The next time Heinreich fired, he had gotten himself much closer to his target and he didn't miss this time. As his plane shuddered from the recoil of his machine cannons, a beautiful stream of tracers arched towards the American plane, and to good effect. "Yes!" he shouted. "I've got you now."

Shards of aluminum fuselage ripped away from the stricken Mustang. Two seconds afterwards the sleek, metallic American fighter burst into flames spewing a mist of volatile fuel, fire and black smoke when its wing root tanks exploded. Inside the Mustang, a twenty-two year old kid from Kansas, whose father had shot down a famous German ace the day before the end of World War I, burned alive in his cockpit. The circle of ironies was complete. Alas, Heinreich, now almost out of fuel and starting his descent towards the Fatherland, would not know of the irony of his kill until many decades later.

"Okay Peter, let's head out," Heinreich called out.

"I'm afraid I won't be able to my friend," Peter replied. "One of the American bomber gunners pierced my fuel tank. I'm going to have to land soon."

Peter's situation chilled Heinreich's blood. "You can't Peter, or you will fall into Soviet hands. We are flying over their tank columns as we speak. If they catch you they will shoot you!"

A second later Peter's right engine flamed out, and Lise and Aunt Lilo, riding in a truck down below, saw Peter's jet begin to plummet. "What can I say Heinreich? I am out of fuel."

Heinreich only then realized how stupid he had been. Before taking off, Peter had begged him to fly towards the American lines to surrender themselves, but Heinreich, ever trying to avenge the deaths of both his father and mother, had insisted they first engage the American bombers before leaving towards the Americans.

"Then I will go down with you mein friend," Heinreich replied.

"Nein!" Peter retorted. "You have to find Lise. I will be fine." Peter's left engine then flamed out and his sinking Me 262 now began to lose altitude like a rock.

"Let me follow you down then. If I give you cover fire, perhaps you can find a car and get away."

Peter declined Heinreich's offer. "Nein. If you shoot at them, all you will do is anger them. Then I truly won't have a prayer's chance. Just leave before you yourself run out of fuel."

Heinreich remained silent for a moment looking down at the Russian column. He was sure he had enough bullets to destroy at least half the trucks and tanks. He lowered his nose and aimed his guns at the leading truck.

On the ground, the Russian soldiers began shooting at his jet. Lise and Aunt Lilo were scared for their lives as the German jet began diving at them. Heinreich put his fingers to the trigger and took aim. As he pulled his trigger, a large explosion rocked his jet. A gaping hole had been ripped through the center of his left wing.

"Pull up and leave you stubborn idiot!" Peter yelled at Heinreich. "If they kill you, then who will come looking for Lise and for me?" Peter asked.

Heinreich, having once again acquired his target, realized the wisdom of Peter's words. As much as he wanted to destroy the Russian column, he pulled away. "I will find Lise and I will find you too Peter. Godspeed mein brother," he called out.

Peter, having picked out a road upon which to land, in turn replied, "Godspeed mein brother," before turning off his radio.

At the exact moment Peter fell into Soviet hands, high in the desert mountains of New Mexico, primitive electro-mechanical computers were toiling away at heaps of lethal calculations that would deal Japan its atomic deathblows. Tyrannosaurus Rex no longer stalked the Earth. 1945 was the time of Colossus and ENIAC, and of the beginning of the end of the ones called the smart mammals who were at the time—except for a prescient few—creating in blissful ignorance the very tools that would lead to their own demise as a species of natural selection. Man-made evolutionary neurobiosilico nanogenetic technologies were founded on the tripartite of controlling the atom based on computer calculations in order to make war machines. These were the progeny of the Manhattan project, as were the so-called mindspace wars that erupted upon humanity less than ten decades later.

ALEX ALANIZ, PH.D.

ARMS RACE

After Heinreich landed his jet on an American occupied airbase, much to the surprise of the ground crew, he was promptly taken prisoner. As a high ranking officer and a jet pilot, Heinreich was classified as a prisoner of high importance and was afforded every professional courtesy. Even during his preliminary interrogation, the Americans offered him coffee, sweat-cakes and chocolate. Peter's fate was not so fortunate. After being beaten to a pulp by the Soviet infantrymen that captured him, causing him to lose three teeth, he was locked into a pin full of German soldiers of all ranks to await his own interrogation and likely execution.

By the end of the week, Heinreich, along with a group of similarly classified important prisoners, was deported via railway to Berlin. After nearly a year since the last time he had been home, he was happy to be returning. When he arrived, he was shocked at how utterly the old city had been wrecked. Most of the buildings he had grown up admiring had collapsed under the extensive Allied fire-bombing campaign. Hardly any structure stood—the few structures that did were riddled with bullet and cannon holes—and all the streets were cratered. Worst of all, Heinreich was horrified when he learned that the Americans and the Soviets had split his home town into two halves, East and West Berlin, and that his family's properties fell on the Soviet side of the dividing line. What concerned Heinreich the most, however, was trying to figure out when he would be freed to search for Lise and her family. He sensed they were nearby.

When the prisoner train arrived in Berlin, Heinreich and two other officers were driven to an old inn that had survived the Allied bombing. He was given a decent room with running water on the fifth floor. A young Army soldier was posted outside his door. Two miles from the inn, on the Soviet side of town, Lise and Aunt Lilo were still being forced to work as translators by their Red Army captors. They had to wear special, orange colored badges, and were not permitted to leave the confines of the heavily guarded Soviet compound.

When Lise could, she spent her free time in the Soviet camp talking to the Germans who were employed as day laborers. She would ask them to put up notes around the neighborhood shops for information about the fate of her family and her husband. These laborers were happy to do so until they learned that Lise and Aunt Lilo were Jews. They then took to spitting on the women.

Once Lise bumped into a woman who had worked as a servant for Frieda. Her heart skipped a beat of hope, only for her spirits to be crushed when she learned that Frieda had died years before, that Minister Boltzmann had committed suicide, and that Heinreich hadn't been seen in town for over a year.

"Heinreich was alive as recently as a year ago," Lise reported to Aunt Lilo later that day.

"It's something to cling onto dear child," Lilo told her niece. "Now if we can find word of your parents and sister."

The minute Dr. Reber had been released from Dachau, he made his way to Auschwitz. The journey, sometimes by foot, sometimes by riding on the back of some truck, took the emaciated doctor nearly three weeks to complete. The closer he got to the place, the worse the stories about it became.

Auschwitz was a death camp he was told. Jews by the tens of thousands were gassed their every year. He tried not believe the awful stories, even when they came from survivors walking their way back to their countries.

When Max finally reached the camp, the Soviet troops posted around the Auschwitz camp area prevented him from entering. "It's too dangerous," the soldiers told him, referring to the stench and infestation of rotting corpses. "But I am a doctor," he explained to them with sign language. Eventually Max was able to convince a Soviet colonel in the medical corps that he could be of service if he were only allowed into the camp. The colonel assented. "Work with Doctor Ivan Pavlovovich, and we will see what we can do for you."

Once Max entered Auschwitz, he was shocked by the magnitude of the atrocities still evident. The stories of the gas chambers were true. He couldn't believe it, but for the fact that the gas chambers were there. Not far from the gas chambers, bodies stacked like cordwood lay rotting by the thousands in huge ditches dug out by Soviet tractors. In other places, there were chambers full of bones that were to be processed and reduced for their phosphor content. Max vomited more than once that day when he thought that Belinda and Olga might be among the piles of dead.

It took Max two weeks working at the place—tending to injured Soviet soldiers with Doctor Pavlovovich—before he was finally was allowed to see the camp's archives. There were volumes upon volumes of records stored in a large warehouse. Documents concerning supply and rations for the SS people who ran the camp, documents concerning payrolls, documents concerning disciplinary actions and so on and so forth, filled row after row of book racks. It took Dr. Reber fully three days before he found the records concerning the camp prisoners themselves. The records, sorted in alphabetical order in massive tomes, were categorized by methods of execution, which were primarily by means of hanging, shooting, and gassing.

Max began with the hangings choosing the letter R, volume 1, 1943-1944. There were so many Rebers listed, he nearly closed the fat book, but he forced himself to press on. With a knot in his throat, he got to the page containing Rebers whose first names began with the letter B. Sure enough there were three Belinda Rebers listed, but they were all of the wrong age. He pressed on to the Olgas. There were nine Olga Rebers, two of under fifteen years of age when they were taken out to the gallows and hung for stealing bread. His Olga was not one of these Olgas.

Max broke down sobbing under the strain. He didn't feel he had the strength to press on to the shootings records. Doctor Pavlovovich volunteered to go through the shooting deaths, which were far more numerous than the hangings. While Doctor Pavlovovich leafed through the pages, Max sat in a corner cringing.

"No. I didn't find neither Belinda your wife nor Olga your daughter. Do you wish me to continue on to the gassings Dr. Reber?" Dr. Pavlovovich asked Max.

Max took a moment to compose himself. "We will do it together please."

The Soviet doctor assented. "Very well."

When Max spotted Belinda's name—she had the right age and address—he froze. Deep inside he had suspected she was dead. Now that it was a fact, he went numb.

Dr. Pavlovovich pointed out another book to Max. "Is this your daughter?"

Max looked at the record. "Yes," he replied.

Belinda and Olga had met their end on earth together in

an ungodly gas chamber. "The bastards killed my girls Ivan. My sweet, sweet girls, they are gone from me. Stolen!"

For the next many minutes Max—seeing Belinda and Olga stepping into the cattle car in his mind's eye, then into the gas chamber without him there to hold them, to bring peace to them—wept, moaned and cursed his way thru the most painful moment of his life. "Why oh why didn't they take me too?" he kept asking while Dr. Pavlovovich kept quietly out of the way.

A week later Max thanked the Soviet colonel and Dr. Pavlovovich for their kindness. "I'm off back to Berlin," he told them, "to look for my firstborn daughter. Thank you Ivan."

Ivan hugged Max. "May God help you find her," he told him.

"Poor bastard," the colonel added after Max was out of earshot.

"I imagine, given all the camps we, the British and the Americans are finding, that Dr. Reber is far from unique," Dr. Pavlovovich concluded. The truth to Ivan's words sickened the colonel.

***

When General Krukov, who was the Soviet general in charge of the administrative division in Berlin, learned of Lise's technical background, that she held a doctoral degree in nuclear physics and had won an important prize for her work from Hitler himself, he passed the information back to Moscow. It took the information about two months to trickle to Sarov, where the Soviets were running a secret atomic bomb project of their own.

When Igor Kurchatov—who was the chief scientist working under Stalin tasked with running that lab—heard about Lise, he immediately picked up the phone and demanded that she not be released. "We need her," he personally told General Krukov. "Don't let her disappear. It would upset me and Comrade Stalin." An old copy of Lise's dissertation, and the data that it contained, was one of the main references being used at his lab.

When Lise was told she was to be ushered off to Russia by a Soviet colonel—she was given no reason—she grew hysterical. "But colonel, I have to find my husband and my family." The colonel gave her his apologies and walked off.

When Lise told her aunt later that day, Aunt Lilo herself fell apart. "They cannot send you away. You are not a war criminal. Why on earth would they want you?"

Lise had a good idea. "To help the Soviets build an atomic bomb and keep up with the Americans."

Aunt Lilo, clinging to her niece, was beside herself with grief. "When will they send you?" she asked with resignation.

Lise, replied through her tears. "This Friday or Saturday, I'm not sure exactly."

"What will become of us?" Aunt Lilo asked.

"I don't know," was all Lise could muster to say.

***

Max Reber was in a cloud of fog when his train rolled in to Berlin on Saturday, 8 September 1945. Dr. Pavlovovich, feeling pity for Max, had given him the necessary funds to travel home by rail. Max was hoping to get word of Lise or Heinreich, if they still lived that is. One thing he knew for certain was that if Heinreich had survived the war, he would be hell-bent on finding Lise. This thought was one of the few ones that comforted him.

It was crowded and noisy when Max stepped off the

train in Berlin's main rail station that afternoon. Soldiers in American, Soviet, and English uniforms mixed among hundreds of civilians waiting for their various connections. As Max walked out of the station, his eyes caught sight of the backside of a woman wearing a dress in the style that Lise liked. The woman was standing by several Soviet soldiers. He let his hopes rise as he walked towards her. The woman had Lise's stature and the same color of skin and hair. When he reached her, with his heart pounding, he tapped the lady on her shoulder. The woman turned around holding a baby in her arms. It was not Lise. "My apologies madam," Dr. Reber told her.

Standing only one meter beyond Max stood Lise escorted by a female Soviet officer. Lise herself had been asked to wear a Soviet officer's uniform to better facilitate her travel into the Soviet Union. Her hair had been shortened accordingly. Max saw her, but disappointed as he was for being foolish enough to think he could so easily run into his daughter, he turned away from Lise. Lise, he told himself, was not in the Soviet military. That young lady could not therefore be his daughter.

Walking towards both Lise and Dr. Reber, Heinreich, under escort—he was being shipped off to the United States to help the Americans design better jet aircraft—thought he saw Lise with his keen fighter pilot's vision. Even if she was wearing a Soviet uniform, it had to be her he thought.

"Sergeant Smith," Heinreich told his escort, "That woman over there may be my lost wife. Would you please allow me to check?"

Sergeant Smith, another squat, barrel-chested Army man, who had grown to like Heinreich's sense of humor over the last couple of months, assented.

"Fine, but we can't miss the train. I've got to get you to

England on time so that you don't miss the ship to the States, or it will be my hide."

It was Lise. Heinreich was sure of it when he was but twenty feet from her. His heart began to race as he closed the distance. "Heinreich!" someone called out from the opposite side. Heinreich turned to see the source of the voice. He was amazed to see that it was Dr. Reber.

"Dr. Reber! Please wait!" Heinreich yelled out. "I think I see Lise over there with the Russians."

Max, catching up to his son-in-law, took Heinreich's arm and detained him. "Nein. Alas not. I though it was her too, but after tapping her shoulder, it was not her."

Heinreich's heart nearly stopped. "Are you sure Dr. Reber?"

Max sighed. "Yes," he stated.

It took a few seconds for Heinreich to accept his disappointed spirits. The woman in the Soviet officer's uniform truly did look like Lise—though it had been years since he had last seen her.

"Well thank God you saw me Dr. Reber," Heinreich finally said after he turned his eyes away from Lise. I am being shipped off to America to work on their jet projects. You need to find Lise for me."

Max nodded yes.

"Are Belinda and Olga fine?" Heinreich continued.

Max's eyes filled with tears. "Nein. They were gassed years ago at Auschwitz. Hans had lied to you and I for years."

Heinreich was stunned. Both men, still clinging to each other by their arms, dropped their heads in sorrow.

"Colonel," Sergeant Smith interrupted using Heinreich's equivalent American rank, "We've got to be going right now."

Heinreich motioned the sergeant for another minute. He

reached into his pocket and pulled out a small notebook. "Dr. Reber, you will be able to reach me at the following address. You must find our Lise. I will send you money and help in any way that I can." When Heinreich finished penning his address, he threw his arms around his father-in-law. "I'm so sorry about your wife and your daughters Dr. Reber." Before letting go of Dr. Reber, Heinreich turned one last time to look for Lise, but the woman in the Soviet uniform was gone.

Lost in her thoughts about never seeing her family or her husband ever again, Lise never noticed Heinreich or her father who had stood so near to her. In Lise's mind only Aunt Lilo remained behind as her last connection to her home and a chance to find her remaining family. Given Aunt Lilo's poor state of health, and that she could hardly walk thanks to her bout with childhood polio, Lise held out little hope. When she took her seat beside her escort and looked into the crowds she saw two men—accompanied by an American soldier—embracing and obviously happy to be seeing each other. "If only I could feel what they are feeling, the joy of reuniting" she thought to herself as the doors to her wagon were closed. A few moments later, with a gentle jerk, the trained pulled its way out of the station.

***

Lise's train ride into Eastern Europe was long, cold and arduous, and all she could think about as she was taken deeper into foreign lands was how completely alone in the world she was. Instead of looking outside at the passing scenery—all there was to see was devastation—Lise concentrated on the basic Russian texts she had been given. Her escort—a very stern, black-haired female major in the Soviet Army—spoke only rudimentary German. Lise thus kept quiet, drowning herself in thoughts of self-pity.

For the first time in her life Lise questioned the value of living. She doubted, however, that she had the fortitude to do anything about her feelings. As for pursuing her interests in nuclear physics, she certainly had no interest in helping the Soviets build atomic bombs. How long the Soviets planned on detaining her was anybody's guess, but her stay in the Soviet Union, she suspected, wasn't going to be short.

When, after ten days of travel by rail, Lise finally reached the train station nearest Sarov, a smallish dilapidated wooden structure with a small café, she was impressed with how large a country the Soviet Union was. Exhausted and starving, she was told by her escort that a car would come for her within the hour. The two women meanwhile entered the small station to refresh themselves and have a bite to eat, boiled potato soup with cheese and vodka. Lise found the vodka horrible, but drank it after she was warned that the water was not potable.

Four hours later, the purported car finally showed up caked in fresh mud. Lise watched as a young, gangly man stepped out and walked over to the station's little café. When he came inside and caught sight of Lise, he recognized her from the start. In perfect German he told her, "Doctor Reber, you are more beautiful in person than your photograph would lead one to believe."

Those were the last words Lise expected to hear. "Excuse me?" she asked expressing her confusion.

"My apologies doctor. My name is Joseph Molotov, one of the junior academicians at the lab. Since I conducted my studies in Austria, and speak a decent German, I have been asked to be your liaison throughout your stay. Everyone is excited by your arrival. We use your work as a reference all the time."

Lise was still taken aback. "I didn't come here of my own free will you know."

Joseph lowered his eyes. "I know," he replied. "I'm sorry. But with your help we can bring stability to the world. Currently the Americans are the only nation capable of fielding atomic weapons. We fear they may decide to do—even succeed at—what the monster Hitler failed to do. We cannot let this situation remain unchallenged."

Lise shrugged her shoulders. "Of course not," she replied. "One nation must respond to the madness of another with mutually equal madness. We," she said switching to English, "will have stability by threat of mutually assured destruction. It's mad."

When Joseph, who understood some English, snickered, Lise wondered if she had said something funny. "What?' she asked him.

"You said, 'Mutually assured destruction is mad,' and it certainly is M A D."

Lise's unintended pun caught on a few years later as MAD based detente.

"We should go," Joseph prompted Lise. "The lab is a long way from here and the roads are nearly impassible this time of the year. Already I got stuck once. That's why I was so late."

***

Sarov was awful. The place was as gray and bleak as any other place Lise had seen rolling across the forbidding Russian tundra, but yet somehow uglier. After checking in past the guard shack, Joseph showed her to her quarters, a smallish, drab room with a bed, a desk, a small sink in a small bathroom, with poor heating, one bare light bulb hanging from the ceiling, and one smallish window looking out into a field of dead trees, two skinny cows and mud as far as the eye could see. "Home sweet home," she remarked.

Joseph shrugged his shoulders. "At least you have a window that faces out. My room faces the sewer."

Lise laughed for the first time since she had left Berlin two weeks earlier. "Thank you Joseph." She liked the young man who, during the long, bumpy car ride to Sarov, had talked her ear off about all his crazy ideas.

Before Joseph stepped out of Lise's room he stopped. "Try to sleep well Dr. Reber. Tomorrow you will meet Kurchatov himself, as well as our group leader Yakov Zeldovich. Yakov is wonderful and brilliant, you will see. Our group is a happy group."

The following morning, several hours before Lise's scheduled meeting with Director Kurchatov, Joseph, looking as if he hadn't slept much, drove Lise around the sprawling, ugly lab complex. There seemed to be no regard for aesthetics or cleanliness. Barrels of toxic chemicals sat everywhere, some of them leaking, and all about milled German prisoners of war shackled together into forced labor units. To his delight, Director Kurchatov would have no problems using labor drawn from the gulags for years to come. From espionage to forced labor, Kurchatov, with Stalin's full blessings, was doing—and would do—whatever was necessary to get the Soviet Union into the atomic bomb race. Already data obtained from spies in the United States as far back as July of that year had propelled his program months ahead of schedule.

"You didn't sleep much Joseph?" Lise asked her escort.

"No. I was supposed to deliver certain calculations to my supervisor Tamm yesterday, but with the car breaking down and us arriving so late, I fell a little behind. Tamm would have done the same thing." Joseph was busy translating the atomic weapons documents that had been ferreted out of the United States from the atomic bomb laboratory in Los Alamos, New Mexico.

Lise understood there was a real sense of urgency to the place, and, though she didn't like the atomic bomb business, she understood the paranoia of the Soviets. She had seen the newspaper photographs of the devastation in Nagasaki and Hiroshima, Japan. No nation could survive against such weapons, and though she would never believe the United States capable of turning evil and trying to dominate the world with its new weapons, it wasn't implausible. She had to admit that, not too many years back, she wouldn't have thought her own Germany capable of doing just that, to try to take control of the world by means of its military might, and the Soviets had paid dearly for not being prepared. More than 20 million of its citizens died. The United States, on the other hand, suffered a much fewer 400,000 deaths with significantly little loss of civilian life.

When Lise met Director Kurchatov in his expansive, but disheveled office, she was taken aback by his looks. He almost had the same beard and piercing eyes as Rasputin, the "Mad Monk" who had contributed to the demise of the Russian Romanov dynasty nearly three decades before. His looks however, belied his mannerisms—they were more like those of a polished politician.

"Welcome Doctor von Reber," he began, but Lise cut him off.

"It is Doctor von Onsager after my husband."

Kurchatov cleared his throat. "Well, here we like to call each other comrades, because that is what we are, you included."

Lise didn't like Kurchatov's lugubrious words. "I wish

to know how long I will be detained here, and I say detained because I did not come of my own volition. I wish to return to my home country as soon as possible. I have a husband and a family to find."

Kurchatov, who had been rifling through some papers, stopped what he was doing and peered straight in Lise's angry eyes. "Well," he began clearing his throat again, "I'm afraid I have sad news to report to you along those lines. We reasoned that you would feel as you do, frustrated, angry and so forth. Once we knew of you, and that we wanted to have you, we began searching for your family to bring them here. To make it short, we found these records—," Kurchatov handed Lise the papers he had found—",of Belinda and Olga Reber. As you will see, they were gassed in the Auschwitz death camp."

Lise couldn't believe her eyes, but before she could react, Kurchatov handed her two more death certificate documents, both of these were carefully forged. "Your father, as you can see, was shot for collaborating with Jews at Dachau. Your husband was shot down the day before the armistice by our glorious air forces." Peter Wohlthat, during a severe beating, had told his captors many things, including the fact that his commanding officer was the Baron Heinreich von Onsager. It didn't take the Soviets too long to realize Heinreich had been captured by the Americans and was shipped off to the United States. The Americans were doing exactly what the Soviets were doing, pillaging through the spoils of war. "I'm sorry, but you will find this place to be your new home, and all of us, comrades, will be your new family."

Lise couldn't take another word and fell apart. She had to be escorted out of Kurchatov's office by a nurse back to her quarters.

***

With Germany defeated, the Americans—like their Soviet counterparts—got very busy trying to prevent the capture of as many of Germany's Nazi rocket scientists and jet engineers before the other side got to them. Heinreich, a colonel in the Luftwaffe, trained as an aeronautical engineer, and one of the original Me-262 test pilots, as well as an accomplished ace fighter pilot, was ushered off to the Army's Wright Field in July of 1945 under Sergeant Smith's escort.

"We'll colonel," Sergeant Smith told Heinreich after they got off their ship in New York City, "welcome to my home. We'll stay here for a day or two, and then start making our way to Wright Army Airfield in Ohio."

The Army, in its infinite wisdom, "drafted" Heinreich the very next day. He went from being an oberst in the Luftwaffe, to a full bird colonel in the United States Army Air Forces with the stroke of a general's pen. Heinreich was going to teach American flyers how to fly captured German jets, and American engineers how to make newer, better American jets.

The first thing Heinreich did when he had the opportunity was to telegram Dr. Reber back in Berlin. "Any news of Lise?" he asked. "Yes," was the reply. "I found Aunt Lilo. She received a message from the Soviet Army. Lise died in a rail accident last week." Heinreich went numb when he read the telegram again and again, as if, perhaps, he wasn't reading it correctly.

"She died last week?" he asked himself aloud. "She had been alive all this time, only to die now? Where was she? How did it happen? Could this be a mistake?" Heinreich was filled with questions. He fired off another telegram back to Berlin. "Are you absolutely sure Dr. Reber?" The reply came back the next day. "Yes. Very sad for you my dear boy."

Sergeant Smith could tell something was up with Heinreich the day he received Max's confirmation. "Colonel, are you okay?"

Heinreich took a moment to reply. "No sergeant. Mein wife is dead. Last time I saw her was six years ago. Now I will never see her again."

Sergeant Smith sighed. "Sorry sir." After the two Army men boarded the train, Heinreich locked himself into his sleeper car, ripped his mother's bible to shreds and drunk himself to sleep with Lise's picture in his hand.

The first impression which struck Heinreich while he and Sergeant Smith traveled deeper into America, was how little the enormous country had changed from when he had last visited with Frieda as a young boy. Heinreich wasn't thinking about the buildings, new models of cars, or clothing fashions, and such ephemeral things. Lots of things of this nature had changed. He was thinking about how unscathed the country looked. Europe, even England, was a wrecked mess of toppled buildings and cratered roads, but not America. Everywhere he looked, the people thrived. The power of the American economy, protected by oceans of distance, shocked him. Whereas Germany had lacked basic industrial necessities such as oil, and metals to make engine parts and propellers, and essentials like food, America was brimming with all she needed. "Little wonder we lost the war," he told Sergeant Smith once as they made their way out to Ohio.

The first thing Heinreich noticed about Wright Army Airfield was how expansive and orderly it was. Everything in it was new and in working order. As for the town of Dayton, Ohio, where the Army base was located, it was quiet and simple, with a nice downtown area—good for kids. It was to be Heinreich's new home until the Army told him otherwise.

By the end of his first week in Dayton, Heinreich picked a small home to rent near the base. It had a small yard and a nice view of a nearby park where children gathered to play. Lise would have liked the place he thought, even if it was small—it had only two bedrooms, one common bathroom, a small kitchen near the dining room, a little office and a living room—a far cry from his home in Berlin. As a colonel, his pay was plenty high for more luxurious lodgings, but alone as he was, he had no need for anything fancy. The day after he rented the house, he bought himself a new car, a basic Chevrolet. Then he went out to buy furniture. He found it difficult not to break down several times as he picked out items he thought Lise would have liked. Not a day went by that he didn't think of her, or a night in which he didn't give her photograph, placed on a nightstand beside his bed, a goodnight kiss.

**

Not long after his arrival at Wright Airfield—the Americans were in a rush to get things going—Heinreich met then Captain Burt "Bone Crusher" Pratt and his young bride, Karen, who was pregnant at the time. With flying being in their blood, the two young men, having recently been enemies notwithstanding, hit it off. Bone Crusher, a student in the Army's new test pilot program, holding a degree in mechanical engineering, and Heinreich an instructor with the equivalent of a masters degree in aeronautical engineering, simply couldn't get enough when it came to flying. For Heinreich, burying himself in flying was also a way for him to deal with his losses and loneliness.

After weeks of ground school training in the Me 262, Bone Crusher finally broke down one day and interrupted Heinreich's class. "Colonel, when the hell are we going to fly one of those damned jets?" Bone Crusher broke in while Heinreich was in the middle of an equation up at the chalk board.

Heinreich turned around and faced Bone Crusher. "Tomorrow captain. Unless, perhaps, you don't wish to be the first volunteer." Everyone in the class burst out laughing.

The next morning was filled with blue skies as far as the eye could see. Colonel von Onsager drove over to the Pratt household and had breakfast with Karen, Bone Crusher and baby Reginald. Karen was happy because Bone Crusher was happy, and Bone Crusher was happy because he was about to pilot his first German built jet, the Me 262, which was by far the best jet airplane flown in the war. Karen had only one question for Heinreich. "Colonel, how is Bone Crusher going to read the German markings?"

Heinreich and Bone Crusher got a good laugh out of this. "The markings have been changed to English, dear" Bone Crusher replied.

"And I wouldn't worry," Heinreich added. "Your Captain Pratt has learned that bird inside and out better than anyone else. I assure you he is ready Frau Pratt."

"Well then eat up boys," Karen told them.

With over 1,900 hours flying time, Captain Pratt was indeed ready to fly his first jet. He was a tough, smart and intuitive young man. During the war, he was credited with shooting down twelve planes, including an Me 262, making him a double ace. Like only one other pilot, he had been shot down over France, joined the Maquis resistance, and made his way back to England while evading German soldiers in the French Pyrenees, all so that he could get back into the air war again contrary to policy. Pilots who had been rescued by the Maquis and returned to England were supposed to be shipped back home, lest they got themselves shot down again and thus compromise Maquis identities and information.

Bone Crusher Pratt, above all else, was humble. His

instructor, Colonel von Onsager, had shot down over a hundred Allied airplanes in his time, a stunning number considering that Heinreich had spent half of the war flying as a non-combat test pilot. German combat pilots, unlike Allied pilots, had to fight their way through the entire war. That's how the best of the best pilots of the German Air Force racked up hundreds of kills. Several of those pilots managed to shoot down over 200 Allied planes, and Erich Hartmann destroyed 352. Bone Crusher knew he had a lot to learn from Heinreich.

As soon as Bone Crusher and Heinreich had blasted off into the air, they climbed to about 15,000 feet and played around with the jets so that Bone Crusher could get a feel for the bird. "The flying characteristics of the Me 262 are great, colonel," Captain Pratt called out. Then, without giving Heinreich a warning, Bone Crusher broke off from Heinreich's jet and returned to the airfield. "I think I see some P-51s taking off colonel. Let's go play with 'em."

Heinreich agreed.

Bone Crusher called out to the P-51s over the radio, "Hey you boys down there. How's about a game of aerial combat?"

One of the pilots of the four P-51s called back. "We're precision test pilots, not a bunch of dumb acrobatics, show boating monkeys."

Colonel Heinreich then broke in, exaggerating his German accent. "Afraid of the Luftwaffe boys?"

That got their goat, and the fight was on, two jets against four piston fighters. When it was all said and done, the four P-51s had been flamed several times over during the mock dog fighting. Heinreich and Bone Crusher were unstoppable. "See you at the officer's club boys," Bone Crusher called out to the P-51 pilots as he broke off his last attack. "You precision test pilots owe us a couple of beers."

The rest of that summer and well into the fall, both Bone Crusher and Heinreich flew six, sometimes eight hours a day after all their respective ground school responsibilities were met. They flew everything Wright Airfield had, including most of the captured German and Japanese fighter planes, and even the first American prototype jet fighter, the Bell P-59 which had been secretly tested in California's Mojave Desert. It was very underpowered and tricky to fly. The German Me 262 was clearly its superior.

"We've got to make 'em better than that," Bone Crusher remarked to Heinreich afterwards.

"We will. Trust me. We will," Heinreich replied knowingly. Heinreich was taking to his new country, "the country of mutts," and working on designing secret, advanced jets for the Army. America was going places, and he was doing his part.

It was Bone Crusher's intuitive feel for equipment, and a little tap on the shoulder from Heinreich, that got the attention of Colonel Bill Anderson, who was to head the flight test division.

"How would you boys like to go to the desert and help us build and fly the world's first supersonic machines?" Colonel Anderson asked Captain Pratt and Colonel von Onsager at the officer's club one hot and dry August afternoon.

Heinreich and Burt just looked at each other.

"When can we go sir?" Burt asked about two seconds later.

"How's 'bout a week?"

A whole week later, Bone Crusher, Karen, their son Reginald, baby Ann, and Heinreich found themselves out in the middle of nowhere at Muroc Air Base in the Mojave Desert.

Conditions for the wife and kids were crude, but Karen stood by her husband, and Heinreich admired that in Karen. Muroc Airfield at the time was 28,000 acres of shrub, lakebeds, Joshua trees, and a few hangars and buildings shimmering in the merciless sun. The nearest town was twenty miles away. The wind never stopped howling, and the desolation took Heinreich's breath away, kind of like Sarov's desolate tundra took Lise's breath away.

***

On 18 September 1947, President Truman signed the United States Air Force into existence. On 14 October 1947, Air Force Captain Chuck Yeager, with badly busted ribs—Chuck had fallen off a horse riding the night before with his wife out to Pancho Barnes Fly Inn Ranch—climbed into the Bell X-1 rocket plane from the bomb bay door of the B-29 that was carrying the X-1 up at 8,000 feet.

A whole hell-of-a-lot of engineering work had gone into that bird. The latest addition of equipment it sported—thanks to test pilot's Jackie Ridley's last minute thinking—was a ten inch portion of a broomstick stuck into the hatch handle, so that Chuck could close it with his good hand. Once Chuck was secure inside the Bell X-1, the B-29 bomber dived to pick up speed and started a countdown from 10 seconds. Heinreich and Bone Crusher, riding in the bomb bay, saw Chuck disconnect the X-1 from the bomber. "He's off," Bone Crusher called out over the big bomber's intercom system as he crossed his fingers. If everything went right, Chuck would go into the history books a hero. If things didn't go right, Chuck would be little more than a footnote in history, and it would have been Heinreich's fault. He was the one who had signed off the X-1 as being ready to fly past Mach 1. Bone Crusher, who would have been the next pilot to try, just crossed his fingers and hoped for the best.

After dropping away from the B-29, Yeager quickly lit off all the engine chambers and climbed on out to 35,000 feet. Once there, he turned off two of the chambers. He continued climbing out to 42,000 feet, then leveled off. Once in level flight, he re-lit the 3rd chamber, and at Mach 0.92 he encountered the usual buffeting the engineers had not been able to get rid of. The Machmeter climbed to Mach 0.97 when all of the sudden the Machmeter fluctuated off the scale (the meter was only calibrated to Mach 1.0, that's how much faith there was in the project at that time in being able to exceed Mach 1.0).

Chuck Yeager said, "Hey Ridley, that Machmeter is acting screwy. It just went off the scale on me."

Jack Ridley retorted, "Son, you is imagining things."

"Must be," Chuck replied.

In point of fact, as confirmed by radar, Chuck had busted through the sound barrier flying 1.07 Mach that day. A lot of people on the ground—listening to their first ever sonic boom—thought that Chuck had bought the farm. "Oh my God," they yelled, "Chuck's done blown himself up!" When everyone realized what had happened, that Chuck was okay, and that he had broken through the sound barrier, one joker remarked, "I'm glad the captain is fine, and that Bell X-1 too. It's a six-million dollar bird. He's only worth six grand a year in salary." Needless to say, Heinreich, Bone Crusher and many other did some raucous celebrating that night in honor of Captain Yeager's historic feat.

A few days later Lise saw a photo of Captain Yeager standing by the Bell X-1. Standing next to him, but cut out so that only his arm showed through, was Colonel Heinreich von

Onsager, USAF. "Heinreich would have loved to have flown that rocket plane," she thought to herself before pitching the paper into a waste basket.

***

The Soviets knew how to enter the history books as well. Much to the consternation of the United States, they exploded their first atomic bomb on 29 August 1949.

Lise was standing side by side with Joseph and graduate student Andrei Sakharov—the famous 1975 winner of the Nobel Peace Prize and future Soviet dissident—when the bomb went off. She thought of Oppenheimer's quote from the Bhagavad Gita, "I am become death, the destroyer of worlds." She too, now, had become death, and felt sad for the world—and sick inside. All this she entered into her diary.

Later that evening Joseph came to Lise's temporary quarters at the test site with two bottles of vodka. "One for you and one for me," he told her after she opened the door. "I figure you would like to celebrate our becoming bastards."

Lise let him in. Since she had arrived in the Soviet Union in 1945, Joseph had looked after her, and, she suspected, he had fallen in love with her. She had done her best not to encourage things, holding out hope against hope that Heinreich was not dead, and that she would yet be with him. But seeing the atomic mushroom cloud made her realize that the world went on. The world had been—past tense—involved in a conflagration of Hitler's making. Now the world was an atomic world, armed with jet airplanes and computers, and rudimentary missiles capable of delivering atomic death from above. Heinreich was a part of Hitler's conflagration. She was a part of the new madness.

Lise opened her bottle and took a swig—she had already

worked her way through half a bottle before Joseph had come. "Tonight is a night to forget the past Joseph," she told her companion while she wiped her mouth clean. "My Heinreich is dead isn't he? Do you know I last saw him in 1939 when I fled to Paris?" she continued asking rhetorical questions.

Joseph looked at her with downcast eyes and shrugged his shoulders sympathetically. "I'm sorry."

Lise took another swig of Vodka. "Don't be Joseph. It's not your fault. Now drink up, for tomorrow there may be no more tomorrows."

Joseph took a hardy swig of his own bottle of Vodka and dragged a wooden chair to a window. He pulled apart the bland curtains and took a seat. "The stars will be out soon," he said after he sat down.

Lise, walking up behind Joseph ran her hands through his hair for the first time. "Do you know I haven't mourned Heinreich's death? Would you lay with me and hold me while I cry?"

It didn't take long for Lise to fall asleep in Joseph's arms after the two of them had gone to Lise's bed. That whole night long Joseph remained awake through the wee hours of the night. He did love Lise, and he wanted her to return his love.

When Lise awoke at the crack of dawn, Joseph's first words to her were, "I love you." Lise replied that she knew, "but I can't dear Joseph. I can't. I don't think I will ever be able to love again."

With the awesome detonation of the Soviet bomb, the world entered the atomic arms race. Heinreich, and the rest of America read about it with incredulity in the papers the following day. Three months later, in response, the Americans, thanks in great part to Edward Teller's pushing, decided to proceed to develop hydrogen bombs, bombs thousands of times more powerful than the ones that had flattened cities in Japan. In the Soviet Union, it was Andrei Sakharov who began pushing for the Soviets to develop their own hydrogen bombs. By 1948 Sakharov had come up with his "layer cake" idea for making these new super bombs work, temporarily surpassing the American efforts.

***

In 1949 Heinreich met twenty-one year old Donna Dearborne, or rather, Donna Dearborne, wearing her favorite red dancing dress, introduced herself to him at the officers' club one Friday night.

"Hey colonel, do you swing?" were the curvy blonde's first words.

Heinreich hadn't danced since ten years before with Lise at the Augsburg's officer's club. "Nein, but thank you."

Donna, the new base commander's only daughter, wasn't the kind of girl who took no for an answer.

"Oh, so you're a German boy? Show me how they dance back home—please!"

Just then someone put a nickel in the jukebox and Louis Prima's 'Sing, Sing, Sing' began to play. Before Heinreich could get to his feet properly, he found himself being yanked out to the dance floor feeling very self-conscious. Heinreich hadn't held a woman in over ten years. Holding Donna's hands in his hands for the twirls, looking at her gracile figure hardly veiled beneath her dress, at her cheery green eyes, at her button nose, and at her giddy, bewitching smiles, was fun.

"Thank you for the dance," Heinreich told her when the music died off.

"Don't think you're getting away that easy colonel. Would you buy a girl a drink?"

Heinreich, ever the gentleman, escorted Donna back to his table and had the waiter bring her a scotch—his favorite drink after a hard day's flying. Heinreich didn't realize that he was being ambushed that night. Ever since her father had taken command of Muroc a few months before, Donna had spotted Heinreich at the change of command party, and had liked what she had seen, a tall, blond, handsome man with well polished European mannerisms and a great sense of life.

Over the months since her father took over Muroc, once she found out that Heinreich and Bone Crusher were best of friends, Donna had learned as much about Heinreich as she could. Most everything she learned came from Karen Pratt who was pregnant with her third child at the time. "He is a widower and a German baron," were some of the first words Karen passed on to Donna.

"Does he have any children?" Donna had then asked.

"No. All I know is that Heinreich is a gentleman, and he needs to meet someone nice. It turns out his first wife was a Jew, but the Nazis didn't figure it out 'till after the war had started. She escaped for a while, but they gassed the poor girl's family in Poland. Colonel von Onsager had thought they were alive. That's how they coerced him to stay in Germany and fight for the Nazis. Shortly after the war ended, he received news that his wife had died in a train wreck."

Hearing this made Donna feel sad for Heinreich. "No one," she told Karen, "should have to go through this kind suffering. No wonder the poor man keeps to himself."

At the end of the evening, Heinreich walked Donna to her car. "Well colonel, I can tell you one thing, seeing as how small Muroc is, I promise you'll be seeing a lot of me." Just before Donna stepped into her car, she gave Heinreich a small peck on the cheek. Lise used to do the same thing he remembered as he watched Donna's car drive away into the darkness.

***

Donna kept her promise. Every Friday night at the officer's club she would make sure to dance with Heinreich. Following Karen's advice, she would also make sure to bake him cookies and given them to him at Sunday services. In this gradual way, thanks in great part to Karen's complicity, things began to develop between Heinreich and Donna.

Karen held several small, intimate dinner parties, and sure enough Donna was always invited to keep Heinreich occupied. These dinner parties led to picnics, shopping trips off base, private dinners, a movie show and so forth.

"So I heard from Captain Pratt that you are a great test pilot," Donna mentioned to Heinreich at one of those dinner parties several months after the first one. "I've never been in one of those rocket planes. Do you think you can arrange to take me up?" she asked him. "I promise I won't tell my daddy if you won't."

Heinreich burst out laughing. "I won't tell your father anything because there won't be anything for me to tell him," he replied.

"I'm serious," Donna insisted. "I'd love to go up in one of those jets. I want to see the Earth from high above." Karen had coached Donna well for that evening.

"You really wish to fly?" Heinreich asked Donna.

"Yes, with you. Please!"

Heinreich—with Bone Crusher winking an eye at him— finally assented. "Fine. I will tell you when, if that is okay with you General Donna."

Donna smiled and gave Heinreich another peck on the cheek. Karen's advice had worked.

A few weeks later, Heinreich snuck Donna—wearing a flight engineer's uniform and helmet—onto the flight deck. He helped her climb into a tandem seat P-80 Shooting Star, and though the crew chief couldn't help noticing the flight engineer had a great coke bottle figure, he didn't say a word to Colonel Heinreich. A few minutes later Heinreich and Donna taxied onto the runway and thundered off into the wild blue.

Donna had a blast while Heinreich took the jet as high as it could go. "It looks like an ant world down there," she said to him as he rolled the P-80 inverted. "From up here, the way we behave down there must look silly," she continued.

"I know what you mean," Heinreich replied. "Now look over there. Do you see the moon?"

The moon was just starting to rise above the horizon, and the afternoon sun lit it up.

"It's beautiful colonel. Do you think we might go there some day?" Donna asked.

Donna's remark made Heinreich think of Hans. "I don't see why not," he concluded.

An hour passed by in this way, with Heinreich and Donna talking about trivial things. It was cozy conversation, not very deep. Being in the sky, being in his element, helped Heinreich relax. He truly enjoyed sharing a slice of his world with the sexy, eager young lady who, he knew, clearly had a crush on him.

After Heinreich and Donna landed, she asked him if he would take her out for a drive to see the sunset. She had packed a picnic basket in advance. "I want to see the stars colonel."

Heinreich, whose feelings for Donna had been growing increasingly mixed over the months, agreed. They drove off to a secluded place, and had dinner on the hood of his Chevrolet.

"Hey, I think I see the first star," Donna said pointing to

a point of light sparkling at the edge of the horizon. "That is Venus, goddess of love," Heinreich corrected her.

"A penny for who spots the next star," Donna challenged Heinreich. The next star just happened to be a shooting star that blew itself apart like a Roman candle. It gave Donna the courage she needed to speak her mind.

Donna took Heinreich's hand and made him look into her eyes. "Heinreich, you've got to let go of her. I'm sure your wife would have wanted you to be happy."

Heinreich, still not ready to hear these words after all the years that had gone by without Lise, cast his eyes skyward and gazed upon the stars above him. "I know," he replied sounding resigned.

Donna reached out and took hold of Heinreich's face. "Look at me!" she told him as she forced him to turn towards her. "I can make you happy if...if you let me."

Heinreich, trying to resist, allowed himself to peer straight into Donna's eyes which were glistening in the starlight. Throughout the months Donna's effervescence had brought him happiness, but sadness as well.

"I don't know if I can let go of Lise—"

Before Heinreich could utter another syllable, Donna had thrown her arms around him, and found his lips in the darkness. That night she made him let go of the past and gave him a future. "You don't have to let go of her. You don't have to stop loving her. You just have to make room for me." Heinreich wept as he made love to Donna.

Before the end of the year, there were wedding bells in Heinreich's life for the second time. Instead of picturesque Capri, the wedding took place at Pancho's, the Muroc Airfield test pilot's favorite (only) real watering hole. The best man was Captain Pratt, and the matron of honor Karen. General

Dearborne, who had hated Germans throughout the war, had grown to respect Colonel von Onsager as a consummate perfectionist and gentleman throughout the year he had worked with the man. He was glad that his daughter was marrying him, and not one of the younger test pilot hotdogs.

General Dearborne and Dr. Reber, who had flown in for the ceremony, hit if off in the old bar. While the young folk danced the night away, they talked about old times and old American movie stars before the war. Aunt Lilo, who had grown much weaker, had stayed behind in Berlin, but had sent her best wishes. Donna's mother, who had divorced General Dearborne while he was away at war, didn't bother to show up.

***

A year after their blissful wedding, Donna gave Heinreich his first son, Karl in 1951, named after Heinreich's father, and in 1952 she gave him his second son Douglas, named after her own father. Heinreich had found happiness again.

By this point in Hans' life—having established himself as a wealthy land owner in Argentina using the loot he had stolen from Jews—Hans had two girls, Johanna and Elizabeth, and his boy, Axel. Axel was already ten years old, and like his father, he was showing every sign of being a prodigy in maths and physics. Einstein's genes ran strong in that boy, his unknown great-grandson. Einstein's lost daughter Constance, stricken with breast cancer, died a year later in 1952, the year the Americans detonated their first hydrogen bomb. She was buried in Argentina, to be forgotten by history until a later date.

***

The Soviets were showing themselves to be formidable adversaries in the cold war. It had only taken them a year after the Americans, this time, to detonate their first hydrogen bomb in August of 1953. Andrei Sakharov, Lise von Onsager, Igor Tamm, Yakov Zeldovich and company, were proving themselves to be as capable as any of the physicists of the United States.

On 4 October 1957, the Soviet Union launched Sputnik, and humanity entered the space race. Now, suddenly and rudely, the people, politicians and military commanders of the United States were pushed into a near panic by a bunch of drunken Tatar barbarians. What was to stop the Soviets from building nuclear weapons complexes in space to rain down nuclear devastation upon the United States in a matter of minutes, the Americans asked themselves. To play catch up was the only answer, and the Congress supplied the bucks.

Heinreich, forty-four years old at the time, was happy. He saw an unprecedented opportunity in the Sputnik craze. "That's where the future is Donna," he told his wife over breakfast as she fed him and the boys hotcakes and eggs with bacon. "I think it's time for us to leave Edwards." Muroc Airfield had been renamed to Edwards Air Force Base years before. "I think," he continued as he put down his newspaper, "California might be a nice place to start ourselves a little company."

Donna, who had heard this kind of talk for years, didn't let herself get excited about her prospects of leaving the desert. "Sure sweetie," she replied while she refreshed Heinreich's empty coffee cup and buttered him a slice of toast. "I think that's a wonderful idea."

Heinreich, after he was able to recover most of his West German assets in 1949, had been thinking about starting a small aerospace company, but so many exciting things were happening all the time at Edwards that he could never find the motivation to leave the place. After Heinreich's participation in designing the Douglas D-558 II Skyrocket, which took men past Mach 2 in November of 1953, he had told Donna over breakfast, "sweetheart, soon there is going to be a strong demand for supersonic transport of civilians. Why don't we go to California and start a company?" Then came the Air Force's idea of building a bird capable of Mach 3, and Heinreich scrapped all his plans to go to California. Unfortunately, the notion of supersonic transport turned out to be years ahead of its time—the Concord didn't fly until the early 1970s. This time though, it turned out Heinreich was serious.

"We're going to help the United States build intercontinental ballistic missiles and spy satellites Donna."

Donna nearly spat out her coffee when Douglas asked, "What's that daddy?" and Heinreich poked his five year old son in the belly with a finger. "It's a rocket that zooms right to your tickle spots!"

By 1959 Heinreich's first company, K2HD (Karl, Karl, Heinreich and Douglas) Industries, with forty employees and three engineers, was supplying critical components to the Air Force, Army and Navy nuclear missile programs, and to the CIA's spy satellite program. It was a lucrative business, and quickly made the von Onsagers millionaires, especially in 1961 when the Soviets launched Senior Lieutenant Yuri Gagarin into space—catching the Americans with their pants down again. Hordes of new contracts came flying to K2HD and the von Onsagers thrived commensurately.

***

In Sarov, Lise, forty-seven years old and still a stunning

woman, was beginning to get worried about the future of mankind. On a rare day when the sun shown through the winter clouds, she took the time to express her concerns to both Andrei and Joseph on one of their long walks through the frozen, snowy tundra. Joseph, who had wed a chemistry technician years back and was raising three little daughters, still remained Lise's closest confident.

"I think mankind may be writing itself out of existence soon, and I don't necessarily mean by nuclear weapons," Lise began with puffs of lazy steam wafting out of her mouth with each syllable.

"What do you mean?" Andrei asked putting on his gloves after securing his snow shoes.

"You are going to think I am insane, but the new field of artificial intelligence scares me. It took billions of years to develop human brains. Now human brains, within a few decades—a century at the most—will make artificial brains so much better and faster than our own. We and the Americans are already building massive military networks to tie all these machines together to control our missiles, soon with satellite-based sensors. What will stop the machines when they realize how stupid we are from ultimately getting rid of us?"

Andrei, a man as prescient as Lise was, had thought along these lines before. He picked up a nice walking stick and shook the snow off of it before replying. "Evolution is not just about a passive natural selection Lise, some kind of simplistic adapting to new conditions lest otherwise one fades away a misshapen misfit. It is also about an active, aggressive, predatory probing for opportunities and niches in the jungle. I think this is what you mean by your question."

Lise agreed with a nod before Sakharov continued. "We are all little cockroaches looking for our niche, trying to order ourselves somewhere, somehow, even at the cost of disordering the world, so long as the advantage is for us as individuals, even if that means making our own replacements or doomsday weapons. If the Americans build a better bomb, we have no choice but to do so in kind, even if it endangers all of us. And they have no choice but to try to build better bombs because they have to assume we are doing the same, which of course, we are doing."

Joseph, broke in with a naïve thought concerning Andrei's comments on evolution. "The only thing evolving, Andrei, is our machinery. Mankind will never change, except very slowly over the course of millions of years—"

"—I am not so sure," Lise interrupted. "Recall what Crick and Watson did eight years ago in discovering the helical structure of DNA. We humans have been reduced to so much genetic software, and I suspect the day will come when we can manipulate this software with viruses perhaps. That is what viruses do after all if I understand my biology. We just need to learn how to make viruses do our bidding in reprogramming our genes to build customized molecular factories."

While Andrei and Joseph digested what Lise had said, the trio walked on crunching a trail of snowshoe footprints into the pristine snow. The brisk air, the snow covered trees, the endless horizon of white, the sparkles of light reflecting off the hoarfrost dancing about their eyes, gave them the impression of being in a fairy tale land.

"Think of Genghis Khan and others of his ilk," Andrei started up again. "When Khan died, so too, rather quickly, did his empire die. The same was true for Hitler. He made his order such that one day his people reaped the disorder of his demise. At the cost of increased global entropy, burnt villages, razed cities, smashed, rotting bodies, a few men, like nature, can obviously spawn periods of extreme, localized self-organization. Think of Alexander's Legions, Napoleon's Grand Armeé, and Hitler's Third Reich. They serve but three such examples. Unlike nature, however, with her 400 million year old sharks, these men could not sustain their creations for very long beyond their graves. Rome fell. France faded after Napoleon died. So too, thankfully, did the Third Reich after Hitler died. Suppose instead that Germany won the war. What if, thanks to a science of genetics controlled by viruses such as Lise has proposed, Hitler could have been made, if not immortal, indefinitely long-lived?"

Lise got a chill in her bones. "What an awful thought Andrei."

Joseph then added his own. "What if, Hitler, given Andrei's scenario, could have been fused to computers to control missiles and robots directly with his new super-mind?"

When Andrei heard this, and saw the stupid face Joseph was putting on, he broke down laughing. "This is all science fantasy. I suppose one such super-being would live on each continent, having swallowed the minds of all its less fortunate inhabitants."

Joseph, adding to Andrei's silliness then added, "Well why not one incredibly bored, lonely super-being for the whole planet Andrei? He or she would then be the ultimate Darwinian winner."

Lise supplied a limiting correction to his idea of a single Darwinian winner in a word. "Dinosaurs! Silly Joseph."

Joseph didn't understand Lise's comment until Andrei clarified it. "The biggest dinosaurs grew so big, that when an enemy bit their tail, they wouldn't learn of the bite until it was too late to save themselves. The signal of the bite would simply take too long to reach their brains in time for these beasts to take the appropriate action. In the same way, the speed of light would limit how big a computer-based Hitler super-being could get and still maintain a coherent self-identity safe from attack."

Joseph scooped up a handful of snow and formed it into a ball. "Well then why not let these super-beings leave the Earth after they suck all its resources dry? Then they could fight things out in the solar system and beyond." When Joseph finished, making Lise and Andrei laugh even harder, he pitched his snowball and hit Lise square on the face. "I have control of Jupiter!" he exclaimed.

"And I have Mars!" Andrei replied whacking Joseph with two loosely packed snowballs.

"And I the Milky Way," Lise added ducking another snowy missile from Joseph while returning fire.

"By the way Lise," Andrei yelled at her, "I've overheard that you've been promoted to division leader."

***

Taking after his mother Frieda, Heinreich took his young family on trips all over the world when his boys were still young. "Learning about other peoples is a good way not to get into terrible wars with them," Heinreich would tell them on these trips.

Throughout the passing years, when Heinreich was at Edwards Air Force Base, and afterwards in California raising a family with Donna, he kept Lise's picture placed on the fireplace mantle. Heinreich wanted Karl and Douglas to grow up knowing who Lise was, and he saw to it that his boys also knew of the awful history of Hitler, the Nazis, their allies and their collaborators. Heinreich wanted Karl and Douglas to grow up thankful that Hitler and his forces had been defeated.

As the years passed, Heinreich could recall Lise's face at his wedding ceremony, or at the Castle Academy, or when they made love, or when they last saw each other in 1939, only with increasing difficulty. If Heinreich could hardly remember Lise, he completely forgot about his best friend Peter Wohlthat. Heinreich was sure the Soviets had shot his faithful wingman dead, as they had shot so many other German soldiers and aviators.

***

Peter Wohlthat certainly met a much different fate after the war than Heinreich did. While Heinreich and his American family thrived in freedom, Peter, along with two million other German soldiers, spent nearly ten years locked away in various Soviet gulags before being released to communist East Germany. If the gulag taught Peter anything, it was to look out for himself, although it took him years to learn this lesson.

The first thing Peter did when he was released into the general population of Lubyanka prison in late 1945—he had been quarantined for three weeks after a severe, and nearly fatal bout of dysentery—was begin to watch the other German prisoners around him carefully. To give himself a better idea as to how the pecking order of the place worked, Peter studied the conduct of his fellow prisoners. Those Germans who maintained their Aryan pride, he duly noted, suffered the worst miseries the gulags had to offer.

Lubyanka prison was filled with endless, dimly lit barracks. It presented an awful way of life for young men who should have had better, brighter futures before them. Each barrack was a long, narrow building, about ten meters wide and twenty-four meters long. Two aisles ran through each barrack, with a row of bunks on each side of each aisle. In one corner of each barrack was a separate small room for the German "Activist" whose job it was to supervise things for the Soviets. These so-called Activists were, in truth, Soviet collaborators that prisoners had to watch out for.

By lights out, the prisoners slept on bunks with straw mattresses that were stacked in three tiers. The bunks were built for two men, but at times over the years, the bunks were so crowded that six men had to sleep in a space built for two, and when one person turned, all of them had to turn. If one person in the middle needed to go to the bathroom, everyone in the bunk would know it.

The latrine was nothing more than a board with twelve holes placed over a large pit that was emptied every two weeks, making the barracks wreak. By day, after the morning announcements, the prisoners were treated to twelve hours of hard labor. Those who couldn't hack the labor—or just got sick—just disappeared, never to return or be heard from again. By night, the exhausted prisoners, fed a bowl of boiled potatoes and stale bread, were treated to propaganda videos extolling the virtues of communism and decrying the evils of capitalism.

At first, whenever a fellow prisoner was down, Peter would do his best to help his comrade, even give up his food ration. Peter wasn't unique in this as the Soviet jailors noticed. A lot of other German soldiers did their best to help out their weaker comrades, and for a few years, the Soviets did nothing to deter or discourage this kind of selflessness.

Eventually the Soviet KGB men—once the prisoners were used to a fairly egalitarian routine—began segregating the prisoner populations by social classes to conduct experiments in social dynamics. German soldiers who were noble blooded were excused from their hard labor duties and given larger food rations, books, and even larger accommodations than those soldiers of baser birth. The rest of the prisoner population was forced not only to continue with their usual labors, but to make up the slack for the lost manpower in nobles. It boiled the blood of these less fortunate men, including Peter, to be treated as second class prisoners. "I saved a baron's behind, and for doing so I get this life?" he started asking himself.

The harder things got for the commoners, and the easier things got for the nobles, the angrier Peter got, especially after his bunk-mate, a quiet man whom Peter liked to play games of poker with, disappeared one day after falling ill. Would he himself fall ill and disappear one day, he wondered.

In this segregated environment, the more time passed, the more Peter learned to hate those noble-blooded prisoners who were given more than he was simply because of the fortune of their birth. By his fourth year at the Lubyanka gulag, Peter began turning his hatred into action, organizing a group of thieves and thugs to take things away from the nobles by force if necessary.

It didn't take long for Peter's keepers to notice this trait in him and in a few other clever, hungry, angry prisoners. The Soviet jailors did their best to encourage Peter and his like to be angrier and more bold by making the lives of the nobles easier and easier. The greater the gradient of "wealth" and "privilege" among the prisoners, the greater the dissension between the classes.

Eventually, Peter, consolidating the prison underground, became the godfather of the Lubyanka prisoner population, and became even more feared among the prisoners than their Soviet jailors. Once when an old Nazi general confronted Peter about some missing boots at the cafeteria, Peter had his thugs stab the old man to death while he strangled the old man from behind with a wet towel. "This isn't the Third Reich you idiot," Peter whispered into the red-faced old man's ear just before his heart gave out. "This is my domain."

Once this happened, and Peter's dominance stuck for a few months, the Soviet KGB experimenters pulled Peter and his senior underground lieutenants out of Lubyanka. "Congratulations," the sixteen men were told by a KGB agent. "Now that you personally understand the evils of a class-based society, you will be trained as spies to defend communism, where every man is an equal comrade. When we are satisfied with your training, we will release you to East Germany to work for us in the Stasi, which is the East German secret police we have set up to control the peoples."

The Stasi, not many years old by this point in East German history, was already much loathed and feared by the citizenry. If you got on their bad side—even if only because a disgruntled neighbor, or a even a jealous relative lied about you—you would be sure to disappear for at least a little while if not longer.

Five years of tough training, indoctrination and deep brainwashing followed for Peter and his band of lieutenants. School topics ranged from hand-to-hand combat to how to best recruit vulnerable young men and women and turn them into informants, even against their own parents if needs be. To help motivate himself, Peter forced himself to keep thinking of Heinreich, the baron who had screwed up his life, especially after the KGB informed Peter that Heinreich was not only alive—as Peter suspected—but thriving in the United States. "I will kill him if I ever get the chance," became Peter's constant thought from that day on. To the KGB's delight, Peter excelled at everything that they threw at him.

***

After his release from his KGB school, Peter, as he had been promised by his KGB trainers, was given a job as a section chief in the main Stasi office of East Berlin. The money he made helped pay for his parent's care. Emma and Treban Wohlthat had lost everything after the war when the communists came and took their farmland away in the name of Stalin in 1945. In a single day, the Wohlthat's were forced at gunpoint off lands that had been in the family for generations, holding on only to those things they could carry, like a box of family photographs, an old Bible, a Holy Icon, but not grandmother's silver. The old Wohlthat couple was living in rat infested, dilapidated housing in the slums of East Berlin when their son was released—within a week he moved his parents to much better lodgings set aside for upper level government functionaries.

The abuse Peter's parents had suffered for nearly a decade in his absence only further increased his hatred of the nobles and of the NATO allies, most especially the Americans. "I should have been here for you mother and father," he would tell his parents before going to work in his high rise office building. "It was not your fault son," both Emma and Treban would reply as Peter would step out of his large apartment.

From his office on the tenth floor of the main Stasi building, a day wouldn't go by that Peter wouldn't use his binoculars to spy upon the wealthy decadence of West Berlin. Up there in his comfortable office, he was far removed from the screams, pleas for mercy and shootings going on in the basements down below. Many people made it into that building. Peter, often with the stroke of a pen, but occasionally using his own to hands to crush skulls with a bat, saw to it that not as many made it out.

Peter eventually married a musician in 1957—Katya was

the daughter of a high ranking Stasi official—and fathered a son and a daughter by her. Not long after his wedding, Peter was given a large promotion to division chief thanks to his father-in-law's connections. Though a lot of that promotion had to do with his father-in-law's influence, a lot of it also had to do with the fact that Peter deserved it. Peter was a tireless worker and an indefatigable proponent of communism. He also ran the best Stasi section in all of East Berlin. The power of his new job got to his head rather quickly, and drove him to attain ever larger ambitions, and by 1961 he was promoted to deputy chief of covert operations.

"Peter, you have got to slow down," Katya would tell her husband after the children were put to bed.

"I cannot dear. There is much work to be done to make communism work. The people, they are so damned indolent. One day they will realized the error of their ways. In the meantime they have to be frightened into playing their roles."

Katya would try to put her arms around Peter's back, kiss him, cajole him into bed.

"I can't Katya," Peter would stop her, removing her arms from his body. "I have too many files to look at tonight. You know that. I've told you hundreds of times already."

Katya, rebuffed, would leave her husband sitting at the dinning table where he would begin working through mounds of paperwork. She would go into her bedroom to weep. Eventually, Katya learned to stop crying and find other distractions at the conservatoire.

In early 1965, the Soviets and the East Germans created a special assassination organization designed to kill leading American engineers, scientists and industrialists by means designed to look natural and above suspicion. Members of the group, if caught, would not be reclaimed by the East Germans, and were expected to swallow a cyanide tablet sewn into their gums with fishing line. Peter joined that covert group not only because the money and privileges were significant, but because he finally had a way to get even with Heinreich. Along with his new assignment, he was given a much larger flat, and his children would be allowed to go to college abroad in the hopes that they would eventually serve as moles.

Once the organization had gathered and collated its first hundred names, Peter was given his first assignment, the Baron Heinreich von Onsager. Peter himself had supplied the name. After all the loyalty Peter had shown Heinreich at great personal risk, like flying Lise out of Germany, the baron—typical of his class—had done nothing to seek him out and help him, or, if not him, help his poor parents at the very least. Instead, while Peter had rotted away a whole decade of his youth in miserable Siberia, Heinreich had gone on with his life, meeting a new woman, and making millions. "For this injustice, Baron Heinreich von Onsager, you will pay with your life," Peter muttered the day his mission was approved.

***

In the early morning of 1 July 1966, thirty-six year old Donna, wearing short pants and a light summer blouse, hurried the boys into the back seat of the family Cadillac. The brighter stars were still visible overhead, but the sun would soon be rising.

"Come on Douglas and Karl, stop fighting. Your father wants to get started early. It's a long drive and we need to get past the morning rush hour."

It was Friday, Heinreich had taken the day off, and the von Onsager family was going on a camping trip into the mountains just on the other side of Los Angeles, where there were several lakes high up at 10,000 feet ideal for fishing.

Heinreich came out of the garage carrying the last of the boxes filled with gear. "Is everyone ready?" he asked as he stuffed the trunk.

"Just about dear," Donna replied giving her husband a quick kiss.

"The tent, the sleeping bags, the rods, I think everything is ready mein gorgeous blonde," Heinreich announced as he closed the trunk.

A block down the street, sitting in a rented blue Buick sedan, was Peter Wohlthat wearing a gray suit and a tweed hat for cover. He was spying upon the von Onsagers with a pair of small binoculars. Beside him rested a high powered semi-automatic hunting rifle with a scope and several boxes of ammunition. After Heinreich shut the trunk, Peter saw the baron spank Donna's derrière behind the car, and her slapping his errant hand away. She and he, laughing, wrestled for a few seconds before Heinreich pinned Donna and started planting kisses on her cheeks. "They're happy," he thought as he lit his third cigarette of the morning. "Soon they will all be dead, and I will have myself another nice promotion."

The drive through the Los Angeles highway system went smoothly. The von Onsagers managed to beat the rush hour as Heinreich had wanted. Heinreich, too busy talking to Donna and his teenaged boys, never noticed the blue Buick tailing him several cars behind.

"I say American football isn't going anywhere soon," Heinreich told Karl.

"You're crazy dad. You just like that pansy European football soccer game."

Douglas then piped in, "yeah dad, soccer is dumb. You'll never see kids playing it in America."

Heinreich chuckled. "Of course you will, just give it time."

Donna, putting on her lipstick and brushing her hair, ignored the boys.

"What's so special about American football? The players have to wear pads. They're not real men," Heinreich continued teasing his boys.

Both Douglas and Karl guffawed. "Neither are those guys kicking a ball around like a bunch of girls."

About an hour later the von Onsagers entered the road leading up to the mountain pass. The campgrounds laid about twenty miles ahead. When Peter passed the von Onsagers' Cadillac, no one paid any attention to him. Several miles ahead Peter found a place with a sharp curve at the bottom of a steep grade carved into a sheer wall of mountain rock. It was a perfect place he thought as he hid his car behind some trees and prepared his weapon's scope as rapidly as he could. He knew the von Onsager car would be coming along soon.

When the metallic gray car appeared rounding the curve high above him, Peter was in perfect position. He took aim and waited until the Cadillac was near the bottom of the steep grade just above the sharp curve down below. He fired. Everyone in the car heard the shot, but no one had time to think about it before the Cadillac started to slide out of control. Peter had shot out the left front tire, and there would be no way to stop the massive vehicle from running off the curve and crashing into the sheer wall of rock situated ten feet beyond the lip the of the road. "The crash should kill a few of them," Peter thought as he watched the event unfold before him with a cold fascination.

It was slow motion pandemonium inside the von Onsager's car. Bits of window and windshield glass went shooting into the air the moment the Cadillac slammed into and bounced off the sheer wall. Peter shot out the other front tire after the car bounced away, sending the car tumbling over its roof and front end several times around. Inside the car, in the days before seatbelt use was made mandatory, bodies bounced around like rag dolls. One of the bodies, bloody and decapitated, was eventually tossed out of the back windshield, landing motionless on a clearing.

The crash ended a few seconds later as the Cadillac, spewing steam, came to a stop upside down. Peter pulled out his pistol and ran towards the wreckage. As he passed the headless body, Peter noticed that it belonged to one of the boys. Inside the car, the other boy, with a mess of blood over his face was moaning. Donna was dead. Her head was twisted all the way around backwards, and a chunk of broken off windshield had lacerated her neck. Heinreich lay motionless on Donna's lap, bleeding profusely from the head. Peter was sure he was dead, but he was going to shoot him nonetheless just to make sure. That's when a truck full of highway workers drove by and stopped to help. Before they could get out, Peter started yelling at them. "These people need help! I'm going to get help at Portersville." He ran back up the hill before any of the men could get a good look at him, got in his rented Buick, and drove off feeling no remorse. "Even if Heinreich lives," he thought, "I've ruined his life. I will now make major general in the Stasi and really start to make things change in East Berlin."

***

Two days after the accident, Douglas awoke from the coma he had been in. Whereas Douglas became paralyzed from the hips down when his spine broke, Heinreich only suffered a broken arm, a broken collar bone, and several lacerations to his head and face.

"With these casts on my legs, I guess I won't be playing football anytime soon," were Douglas's first brave words to his father in the recovery ward after Heinreich had been summoned from a nearby waiting room. Douglas had not yet been told of his paralysis. Heinreich had asked the doctors to allow him the grim job.

"I'm so sorry son," Heinreich replied.

"Can you tell me what happened dad? Are mom and Karl okay?"

Heinreich sat down beside Douglas and took his boy's hand. "We had a blowout, a stupid blowout. We must of hit something on the road." This was a half truth. The tires were brand new, and according to police forensics, someone had shot the front tires out. Heinreich would wait for the appropriate time to pass this information on to his son. It wasn't the time to scare Douglas. The boy had a lot of other things he would have to deal with first.

"Mom and Karl?" Douglas asked again.

Heinreich tried to keep a steady voice. "They...they did not make it son. I'm sorry. I've been holding off the funeral services hoping to see you come out of your coma. The doctors tell me you are paralyzed from the waist down. I am so sorry mein son."

Douglas didn't say a word to his father as his father covered his face with his free hand.

Dr. Reber, who hadn't seen Heinreich in nearly three years, flew to California for the funeral services—the doctor looked very aged. Heinreich was glad to have Max help with things. He needed all the support he could get. After Aunt Lilo had passed away, Dr. Reber kept himself busy working in South America. He spent part of his time treating orphans, and part of his time trying to track down Nazi war criminals by means of bank records.

A few days after the funeral services were over, Heinreich would have killed himself if it hadn't been for Douglas having survived, or for Dr. Reber's support. "If Dr. Reber could make it after all that happened to him," Heinreich would think to himself, "then so too can I. Besides, I have to. My son's life has been wrecked, and I need to be there for him."

Within a few weeks, while Douglas learned how to get around in an electric wheelchair, Heinreich sold his home. It contained too many memories for both him and his son.

***

As the United States embroiled itself in Southeast Asia, and things grew messy domestically, Heinreich grew into a very cautious man. Whereas before the "accident" he had hidden his wealth and insisted on raising his boys in a middle class neighborhood, Heinreich purchased a new home in an exclusive neighborhood in the foothills of Los Angeles. To get into to it, residents had to get past a guard shack, and the streets were patrolled by private security forces 24/7. To make life as easy as possible for Douglas, Heinreich had every modern convenience put into the new house, and hired his son a full time nursing assistant until Douglas could figure out how to do the essential things for himself.

Heinreich's new business, branching into communication, began to flourish by the end of the year, and by the end of 1968, Heinreich became one of the few Americans worth over ten million dollars. The more money he made, however, the more he shut both himself and his son from the world at large.

Within his exclusive neighborhood, Heinreich eventually built himself and Douglas a new mansion protected by his own private armed guards, dogs and short-circuit television monitors. Douglas didn't like the changes in his father's character, but it took the young man a few years before he finally confronted his father.

The confrontation—if you could call it that—happened on a trip back to a place in Germany where Heinreich had once lived as a boy. Dr. Reber, while tracking down Nazi war criminals, had died in his sleep in Buenos Aires. Heinreich, taking charge of matters, made arrangement to have the old man's body transported back to Berlin to be buried next to Aunt Lilo.

"Max was a good man in many ways," Heinreich told the small group which came to pay their respects. "He was a loving, dedicated husband. I remember the way his wife Belinda used to look at him, with such glowing affection. I remember the way his two little ones Nanerl and Olga used to cling to him as he played with them. I remember how my own beloved first wife, Lise, respected and adored Max. When the Nazis took medicine away from the Jews, he, at great risk, did what he could to help his fellow human beings." Heinreich paused to wipe his eyes before resuming his eulogy. "Fate was cruel to him. Like Job, he lost everything, but like Job, the man persevered. After the war, he dedicated his life to tracking Nazi war criminals, always making sure though, to find time to treat the needy. Max wanted to make sure that everyone got their due. Thanks to him, seven of these war criminals had their day in court. Thanks to him, thousand of lives were saved. To me, Dr. Max Reber was like my father--like I imagined the father I never met, strong, loving, a great teacher, someone who would always put others before himself...."

After the small, private funeral was over, Heinreich took Douglas on a driving tour of places that had meant something to him while he was a boy, places that had survived the war.

"The old castle that had belonged to mein father and

grandfather is about an hour away," Heinreich informed his son. "Do you need me to stop the car before we go Douglas?"

Douglas, who was busy looking at the impressive silhouettes of the Alps they were approaching, motioned no. "I'm fine dad. Let's keep going."

"I think you're going to like the place that mein father grew up in," Heinreich continued. "It didn't do too well during the war, but most of it remains standing." Heinreich had been back to the old, defunct Castle Academy several times before. It was Douglas' first visit however.

When Heinreich and Douglas got to the old castle, still pockmarked with bullet and cannon holes from a battle fought against the advancing Soviets, Heinreich noticed it was overrun with more vegetation than ever before, but it didn't worry him. Heinreich figured that with about two-million dollars he could restore it to its old glory and open up a new academy for international studies—which is what Frieda had originally wanted to do with the place.

After parking the car near the entrance of the old castle, Heinreich stepped out and helped Douglas get into his electric wheel chair. It was sunny outside, the air was fresh, and everything was in bloom. Life had gone on.

"Wow dad! You went to school in this crazy place?"

Heinreich smiled and nodded yes. "Let's go take a look inside shan't we? I want to tell you about some old ghosts that will always live in it for me." Once Heinreich got Douglas inside and got his bearings, he wheeled his son to where he had first met Lise and Hans. Heinreich took a seat there on a pile of rubble and started telling his son some old stories.

"As you know, I was a young boy when I met the girl who was to be my first wife, Lise Reber. She was pretty and smart and funny. Of course, I had no idea she was going to be my wife at that point. All I knew was that I wanted to be near her."

While Heinreich talked his way from the past to the days when he met Donna at Muroc and took her on a jet ride, Douglas closed his eyes and tried to soak in every word that issued from Heinreich's mouth. There were times for the young man when he had difficulty holding back his tears, and times when it was hard for him to stop laughing, like when he learned that his father had parachuted to his first wedding and fallen into a muddy flowerbed after Peter had "dive bombed" the wedding guests. "Someday dad, you should put your story on paper," Douglas suggested after his father paused to reflect on his life.

While Heinreich took a moment to think about how much had happened since the moment he had stood at the same spot in 1926, Douglas gathered his strength. "Dad, I need to tell you something."

Heinreich turned and looked at his son as only a father can.

"Dad...you can't keep me locked up behind closed doors all of my life."

Heinreich smiled. "I don't do that son. At least I try not to," he replied. "Back home you have a movie theatre and a pinball arcade built into our house, you have your school friends, an ice cream shop you can get to with your electric wheel chair, and—"

"You know what I mean dad," Douglas broke in. "Soon I'll be going to college. Just next year. You're going to have to learn to let go of me. I'll be alright. I want to know that you will be alright with that."

Heinreich rested his hand on Douglas' knee. He knew that Douglas was right about his growing up, and was touched by Douglas's concern for him. "Douglas, I think you can understand that you are the only person I have left in the whole world that means anything to me. I've lost everyone, mein father, mein step father, mein mother and mein first and second wives, and little Karl. I hope you can understand that I'm scared of loosing you too. As you know, someone did try to kill us all a few years back, and they nearly succeeded."

Douglas understood his father's point of view, but it was time for his father to start letting go. "Dad," Douglas insisted, "you can't protect me forever. Whoever shot the tires was probably a nut picking out a random target. We were the ones he chose. That's all."

Heinreich wished he could be sure of that. The police themselves never found any more than two bullet casings and some tire prints.

"Is that it over there dad--the glider field?" Douglas asked as Heinreich took his son on a tour outside the ruined castle.

"Yes. That was the old Castle Academy Glider Club area. When it had belonged to mein father, he had used it to fly large glider kites. Around these parts, by the way, mein father was known as the Baron Karl von Onsager, and mother would often say to me that he was a good man. I'm certain he would have been very proud of you Douglas, as I am."

If there was one thing Douglas never doubted, it was that his father loved him. No matter how busy Heinreich's life got with his companies, he always seemed to find the time for his son.

"Do you ever wonder father?" Douglas asked, "what would have happened if the war had never happened?"

Heinreich, taking in the sight of the valley down below, and then back at the old castle from this new perspective, took a moment to reply. He noticed that the back side of the castle was more severely damaged by Soviet tank fire than the other side was. "Yes. Often. I would not have met your mother for one, and I wouldn't have you in mein life, nor would we have had Karl. Though they are gone, I'm glad we had them for at least a while."

Douglas agreed with this sentiment. "But it will never stop feeling unfair will it dad?"

"No mein son, it won't. I still miss Lise, Donna, Karl and mein mother and stepfather everyday of my life. You will miss your mother and your brother 'till the last of your days." As for Heinreich's thoughts of Lise and her family, dying terrible deaths at the hands of the Nazis in one of their concentration camps, those he kept to himself out of guilt.

Halfway around the world, living in solitude, Lise too felt guilt for having survived the war while no one else she cared about had lived. In Sarov, with the exception of Andrei, Joseph and their wives and children, it seemed to Lise that she would forever be a foreigner and a Jew in a strange land, valued only for her intelligence. As for her feelings regarding her lost family and her deceased husband, the fighter pilot who had been the love of her life, there was no one she could really talk to beyond the echoes in her diary.

ALEX ALANIZ, PH.D.

RELIGIO-ECONO WAR

***

After attending the School of Business at Harvard University, Douglas decided to go into the business of trading futures in 1974. With his father's help, Douglas started a small, computerized brokerage firm to trade futures at the newly founded Chicago Board of Exchange (CBOE). The 'kid', bright at mathematics, had understood the value of the 1973 Black-Scholes-Merton physics-derived formula for pricing complex derivative instruments across the globe's financial networks. That same year Lise was made director of the weapons grade production program at the Novosibirsk facilities, the first woman to be so named, and she well deserved it. Hans had his first grandson, Paul Fritz, who would turn out to be yet another prodigy, and upon whose discoveries a few decades later, the fate of the whole world would one day rest.

Douglas' business grew along with that of the CBOE's exponentially explosive growth. Thanks to the leveraged instruments he traded, he very rapidly became a wealthy man, far wealthier in fact than his own father. By the time the Berlin Wall fell in November of 1989, despite having lost half his fortune to a costly divorce two years beforehand, Douglas' net worth surpassed one billion dollars. His seventy-seven year old father, couldn't have been more proud of his son than when

Douglas donated ten million dollars for spinal chord injury research.

"Dad," Douglas told his father at the ribbon cutting ceremony at New York Hospital, "someday soon we who have been paralyzed will have computer chips implanted in our brains to help us control our limbs, which, other than being severed from our brains, are otherwise fine."

Heinreich, having seen so many wonders since being born in late 1913, had to agree. "I don't see why not when I think of the primitive horse and buggy world of mein father when he was a boy in the 1880s, and of the primitive planes I flew as a fighter pilot."

Lise retired from her post in 1991, the year the Human Genome project began mapping out the DNA software of humanity in earnest. She applied for a position teaching the history of physics in the Department of Physics at the École Normale in Paris, and was accepted the following year, when sufficient funds were made available by the chair of the Department, Professor Pierre Hilbert Lacombe. Professor Lacombe was very interested in writing a book on the history of the Soviet nuclear weapons program. Lise had worked directly with the likes of Andrei Sakharov, Lev Landau, Igor Tamm and Yakov Zeldovich during the times of Stalin and his purges, and she had kept good memoirs of her life and times behind the Iron Curtain. Lise was exactly what Pierre was looking for, and if everything went according to his plan, he would could have his book out in a matter of a few years.

By 2001 Douglas, the *wunderkind*, as people liked to call him because his father was German, had everything going for him as the world entered the new millennium. He was a billionaire several times over, and had trading offices located all over the world. He was very proud of the fact that he employed over six thousand people worldwide, and that his main trading office—trading mostly in the oil and natural gas futures markets—occupied two floors in the north tower of the World Trade Center. From up there the view of the Manhattan skyline was spectacular, and it made him feel like a prince when he'd look at the vista of ocean and skyscrapers.

On 10 September 2001, Douglas returned to New York City after having concluded a lucrative oil deal in Saudi Arabia. He called his father just to say hello. "You doing alright dad?" Douglas began.

"My cold is just about over son. You know I rarely get sick, so when I do, I become a baby."

Douglas laughed. "No way dad. You're still the same tough, fighter pilot hombre I've always known."

Even though Heinreich was eighty-eight years old, to Douglas, his father—who was still flying aerobatics—was still as tough as nails. "Guess what dad?" Douglas continued as he looked down at the length of Wall Street from his office a thousand feet up. "I think I met a nice girl a few weeks ago. You know how its been for me since the divorce. I just keep busy. But Linda—she plays the French Horn at the Met—seems to be different. She's also wheelchair bound—she's been that way since after a car accident four years ago—and I plan to meet her for lunch tomorrow. I haven't seen her in a week, so I'm kinda of scared and excited at the same time."

Heinreich, now living in Houston so that he could be near NASA and his new company's headquarters, liked what he was hearing. "That sounds great Douglas. You should meet a nice girl. You deserve it."

Douglas appreciated hearing this. "So how are things at SolDynetics dad?"

Heinreich supplied his usual answer. "Hurry up and wait.

You know how it is with government contracts. We should hear from the CIA soon enough whether they want us to build the spysat for them or not."

The following morning Douglas never showed up to his appointed lunch date—no one had to tell Linda why as she wept before the television set in her living room. On 11 September 2001, Douglas and most of the people in his New York firm, were instantaneously killed when American Airlines Flight 11 crashed into the building. All Linda had to do was look outside and see the towers collapsing for herself to know that Douglas had probably not made it.

The news of the terrorist attacks nearly killed Heinreich as he watched the incident on CNN time and time again from his office high in the Transco tower in Houston. "Am I Job?" he asked God after both he and his personal assistant tried to telephone Douglas dozens of times. "Have I have lived eighty-eight years only to see everyone I've ever loved killed?"

After 48 hours passed without word from Douglas, and without any more victims being pulled alive from the rubble of the downed towers, Heinreich knew Douglas had perished. He packed a bag and headed to his private jet parked at Houston Intercontinental. From the air he called the vice-president of operations of his new, five year old company, SolDynetics Inc. He knew that Bob Levanthal was a capable man, and told him to take over things for a while.

"Bob, thank you and everyone for all the flowers. Those people who made it from Douglas' firm are going to need someone to get them back on their feet. You'll know where to reach me if you need me."

From the 67th floor of the Transco tower, Bob told his boss not to worry. "If anything happens Heinreich, I'll let you know first thing." SolDynetics was—after several months—still expecting word on their proposal to build a large spy satellite for the CIA. "And by the way," Bob continued, "our transport plane has just picked up the supplies you asked for, including the computers and satellite links. We should get them to New York within a day."

"Thanks Bob," Heinreich replied as he signed off passing thirty-thousand feet of cumulonimbus buildup.

When Heinreich reached the financial district of New York City, it was pandemonium. Seeing the smoldering rubble of the twin towers for himself enraged him the same way it had when Frieda had been bombed to death.

Within a few hours Heinreich met up with the most senior surviving member of Douglas' trading firm, Jana Friedman, a senior attorney who happened to be at the dentist's getting an emergency root canal when the planes crashed into the towers. She was 36 years old, in shape, and though shaken by the recent events, showing a gritty, New Yorker's determination to get the firm's survivors temporary office space. When she saw Heinreich walking up to her, she immediately recognized the old man from all the photos Douglas had kept of him in his office.

"Hello sir. I'm sorry about your son. I'm glad you're here because we could sure use your help."

Heinreich shook the lady's hand, then spontaneously embraced her.

"I'm having computers and several satellite links shipped from Houston by private transport. We should have your people up and running quickly. How many made it?"

Jana pointed to herself. "Including me, nine others out of two hundred and ninety, but we have others in our Jersey offices that are helping." Jana lost her composure after she said this, and Heinreich, like a good grey-haired grandfather, took her into his arms again.

"I promise you, we are not going to let those bastards bring us down Jana."

***

Lise watched the horror of the September 11, 2001 attacks against the United States on her television set in her flat in Paris near the university. It saddened her to think that there were people as ugly as the Nazis had been still out there in the world. Before it had been Aryan supremacy. Now it was Islamic fundamentalism. What would the next wave of evil be based on, she wondered.

"This attack is going to accelerate biometrics," she told Professor Lacombe on their daily walk along the Seine, "along with the big brother technology to keep watch over humanity."

"Maybe, as I fear, the machines will take over after all," Pierre added.

"Or maybe we will fuse with them as I once told Andrei and Joseph. Have you seen the artificial eyes that company Dobelle developed for sending information directly to the brain? Imagine it. Instead of beaming to the brain what the eyes normally see in front of them via the optic nerve, why not beam what some video camera probe sees under the ocean, or what a lander on Mars sees via radio link?"

Pierre added yet another alternative to the end-of-the-world scenarios that he liked to discuss with Lise. "Or maybe we will still destroy ourselves by means of nuclear or biological weapons long before we ever get to the technology behind your prognostication." Pierre had a good point. After the fall of the Soviet Union, the world had only become more violent and unstable as old ethnic hatreds were now free to erupt unchecked.

Hans, from his hospital bed in Argentina, saw the attacks on television as well. Colon cancer was killing him. He was down to a hundred pounds of skin and bone. "Those terrorists are bastards," he told his aging wife Katya with a shaky voice, not at all considering that he had been one of the world's biggest bastards himself living off the loot of thousands of murdered Jewish families. On 2 January 2002, Hans died not feeling an ounce of remorse for all the people he had tracked down as a senior SS officer to be systematically transferred to Germany's concentration camps for liquidation as part of Hitler's final solution. To his children and his grandchildren, Hans died a saint who had been a wonderful husband and deeply caring man, and were glad that he had lived a long, prosperous and happy life.

***

Heinreich, renting a spacious apartment in a glass tower near Wall Street, kept himself busy for the next year helping Jana rebuild Douglas' trading firm. The more he learned about the Saudi's, their internal corruption, and their two-faced hypocrisy of being allies to the United States while simultaneously funding anti-American terrorists, the angrier Heinreich got. On the good side, Douglas' firm not only got itself back on its feet within a year's time, it tripled its net worth under Jana's leadership, making Heinreich a billionaire atop of the billions he had received from his son's estate. When he was sure that everything was under control, Heinreich packed his bags.

"I have to go back home to Houston," Heinreich informed Jana on 11 April 2003. "Keep up the good work dear."

When Jana heard these words she shook Heinreich's hand and then threw her arms around the old man. "Good luck sir and thank you for everything. Douglas would be so proud of you," she told him while brushing tears off her cheeks. "I'll make damned sure to keep your and your son's company moving on."

Heinreich, wiping away a pair of tears of his own, kissed Jana on the cheek. "You do that, but you make sure that you don't forget to live your own life young lady. There's more to life than futures, like a nice bunch of kids to keep you busy."

When Heinreich returned to hot and muggy Houston, he was a man with a mission. "We've got to find a way to get rid of our oil dependence from those damned Arab bastards, President Bush and his damned family connections notwithstanding," Heinreich told his gray-haired, fifty-five year old vice-president of operations once he was back in his high-rise Houston office.

Bob couldn't agree more, but he also made a good counterpoint to his boss' idea. "Until we Americans stop driving our ever larger SUVs and using up—what?—20 to 25% of the world's energy, we're not going to be able to cut the umbilical cord for another fifty to one-hundred years from now."

Heinreich, joining Bob, who was standing at the window, cast his eyes down to the streets nearly a thousand feet below. They were jammed. "You're right Bob, but I feel there has to be something we can do."

Bob rested his hand on Heinreich's shoulder and said, "As a matter of fact there is. Our chief scientist Charlie Moss and I have been working on a pet project using some of the 301 discretionary funds. It turns out that there's a whole-heck-of-a-lot of Helium-3 on the moon. If we could mine it, use it to fuel fusion reactors, beam the energy in microwave form to collectors orbiting the Earth, and then beam the energy to collector farms on Earth, we could cut the Arabs out for good. A barrel of oil would become as worthless as a barrel of rotting dog piss. Hell, Charlie told me a space-shuttle load of that stuff could power the whole United States for fifty years."

Heinreich was intrigued. "If it can be done, then why the hell is no one doing it Bob? I haven't read anything about it in the papers."

Bob had to control his smile. "Long term risk, maybe twenty years and ten billion dollars just to set up the basic infrastructure," he replied. "Even to me, its just something I work on for fun. Charlie is the guy who has real faith in the idea."

Heinreich shook his head. "Well Bob, you know mein story about as well as anyone might. I figure I've been living on borrowed time since the first time I tangled with enemy bombers over Germany. Do you think I've got another twenty years?"

Bob pat his old boss on the shoulder. "Hell colonel, I think you've at least another twenty more."

Heinreich thought so too. "Well Bob, why don't you set up a meeting with Charlie. I want to talk to him and see if this isn't some stupid wild goose chase. If he can convince us, and a few outside experts, we'll go for it."

The first thing Charlie, Dr. Charles Moss, said was, "Colonel, if you're serious, there is enough of the stuff [Helium-3] up there to power the Earth for thousands of years, and rid the value of the oil resources the Arab fanatics are using to finance their troglodyte vision of life." Bob had coached Charlie's words well.

"If you think we can do it Charlie, then let's do it.

Money—thanks to my son's company, and that Jana running things—will not be a problem. Why don't you put together something on paper and convince Bob and I that you're not as crazy as we think you are."

Charlie, grinning from ear to ear, couldn't believe his ears. "Are you serious? You got it colonel. I'll start putting together preliminary ideas and have them on your desk first thing by Monday morning."

A few weeks after Heinreich had met with his chief scientist at SolDynetics, he launched a small company named Autonomous Lunar Mining (ALM) and put Charlie at its head. That hot, muggy Houston day was one of the happiest days in the fifty year old scientist's life. A few moments after the ribbon cutting ceremony, Heinreich surprised the crowd with his statement at the podium.

"If by birth and my zeal for flying I blindly served darkness under Adolf Hitler and the Nazis, now, as long as I draw breath, I will serve democracy and free markets, and fight to crush the retro-evolutionary brainwashed Islamic fanatics skulking in caves, cowering from the light like cockroaches, preferring the company of fleas, sheep and other smelly, cruel, dark hearted men of little minds over cleanliness and women and civil civilization." The group of seventy scientists and engineers of ALM broke into applause along with a few reporters when Heinreich paused. "And Bob, I now have a reason to live for as long as I possibly can—war against our dependence on oil from the Arabs. I'm going to get a physical tomorrow and will double my current exercise routine."

Bob liked what he was hearing as the small crowd started to applaud their boss. "Would you like a stationary bicycle put into your office like mine? You know you have the room. And

I can get you the name of our family nutritionist," Bob yelled back, making the small crowd laugh. Everyone there knew that Bob had been a gold medal winning Olympian representing the United States in several track and field distance races, and that he still kept himself as fit as he could.

"You do that Bob, and thanks for looking out for me throughout these tough times."

ALEX ALANIZ, PH.D.

IN THE FUTURE...

As the Nazi's would have said, Heinreich was a man of good stock. Ironically, so too was Lise, of the Jewish infestation. They were both long-lived. It was in their genes. Even without the fruits of post-genomic medicine, Heinreich was built to last as much as 110 years, and Lise likely even longer. When Heinreich turned 91 in 2004, he finally broke down and had a pair of cochlear implants sewn into his head—without them he would not have been able to continue flying legally. The implants would be only the first of many bionics to come for Heinreich. Lise, keeping herself busy trying to master the latest developments in particle physics, and enjoying her freedom in Paris, was still doing fine at 90. She loved her constitutionals with Professor Lacombe along the Seine, talking to him about the world past, present and future while tourists and lovers strolled by hand in hand.

The year 2004 was the year a civilian group successfully put a man into suborbital space in SpaceShipOne. By this point in time, Charlie's group of scientists and engineers had come up with a preliminary design to collect and transmit microwave energy to the Earth from the moon. There were only two problems. Practical fusion physics still seemed to be a long way off—perhaps another three decades—and Autonomous Lunar Mining had no rockets to get their hardware into orbit and to the moon in the first place. According to Heinreich's chief scientist, the resolution of the former problem was potentially the easiest of ALM's two main problems: to forget about fusion physics. The other, latter problem was more a matter of money, not of fundamental science or engineering.

Heinreich, after putting off meeting with Charlie for a few weeks so that he could recover from his cochlear implant surgery, finally called his chief scientist for a private meeting to discuss both issues. Charlie liked Heinreich's Transco office up there in the sky. It had a view that stretched from horizon to horizon for miles on end versus his windowless office building in Galveston, near NASA. A lot of the things being worked on in Charlie's nondescript building were too classified for prying eyes, like the satellite spy sensors being assembled for various CIA and Air Force satellites in collaboration with Heinreich's main company, SolDynetics.

"Mein new ears are working great," Heinreich announced while pouring himself and Charlie a glass of tea. "You won't have to speak up so loudly anymore. So how are you doing Charlie? You're looking thinner."

Charlie took his glass of tea and wiped the sweat from his brow. "I'm doing fine sir—inside an air conditioned building that is. It's hell on Earth outside. And yes, I've lost a few pounds. My new girlfriend is trying to whip me into shape."

Heinreich laughed. Since he had known Charlie, Charlie had gone through half a dozen girlfriends. "Good for you Charlie, and as far as it being hell on Earth, I agree. We're certainly learning how to bake the Earth with global warming. So tell me Charlie, people are saying fusion is still decades away. Are there any alternatives?"

Charlie put down his half-empty glass of tea. "Hell sir, sometimes I think fusion power is nothing more than a government boondoggle, and that the guys doing the research are just dragging it out 'till they and their graduate students retire. There are definitely times I think fusion will just never work. But there are alternatives as you suggest. We can collect solar radiation directly from the sun and bypass the whole Helium-3 and fusion idea. The only problem is that the solar collectors will have to be huge given our current conversion efficiencies, and building such structures in space might prove difficult. We might need robot assemblers."

Heinreich scratched his head. "Why not try for both, fusion and solar?" he suggested. "We have the resources. With oil supplies tight and demand high, Jana is making mein son's company's valuation soar, and SolDynetics isn't doing too bad either. Just hire another team to go after the solar power microwave solution."

Charlie just shook his head and smiled. "You do realize colonel, that this is an ideal job you've given me. Whatever I need I get. So unless you fire me, I'll never quit."

Heinreich pat Charlie on the back. "I'll fire you unless you get an experimental solar collector and transmitter in space before I'm dead." Both Heinreich and Charlie had a good laugh over that remark.

"Now as for ALM not having any space faring capabilities, I have several ideas Charlie. One of them is to either buy, or at least get license rights to the technologies being developed by the winners of the ANSARI X-Prize for getting people into space without any government monies. The other is to pour money into the development of a space elevator tower rising 62,000 miles into space. Some people at the University of Houston are telling me that material strengths, thanks to the newly developed carbon nanotubes, are approaching the necessary level to make such a crazy idea practicable. Get some people on that Charlie. Don't let any stone go unturned. We can't afford to waste time."

Charlie finished gulping his tea, sucked out an ice cube, stood up and walked to the glass wall and began crunching the ice. "You have got some view here sir."

Heinreich joined Charlie at the large window. "If our ideas work Charlie, someday your view may be from a few hundred miles up."

Heinreich got himself so busy that year that he didn't notice the publication of Professor Lacombe's book, Memoirs of Soviet Nuclear Scientists, in which Lise played a prominent role. It was published in French for one, and didn't receive much publicity in France or elsewhere. Only a few scholars in various political science programs got notice of the book's release. Some of these scholars picked up the book as a textbook on Cold War politics, while others as a reference book for furthering their own research interests. At any rate, Lise was pleased that her memories of Joseph, Tamm, Andrei Sakharov and others were preserved for future generations.

"They say that those who do not know history are condemned to repeat it," Lise told Professor Lacombe when she received her copy of the book in its final print form. Professor Lacombe didn't agree. "Looking at the Americans, giving up their civil liberties with their Patriot Act, I'm not sure the aphorism is necessarily true."

"Hitler began," he continued, "the same way didn't he? Protecting citizens from evil forces?"

Lise didn't want to go so far. "Hitler did become dictator in the name of protecting German citizens from communist terror, but I don't think the Americans will ever let a president become dictator like President Hindenburg did. It is not in their constitution nor in their beloved Constitution."

***

By early 2012, Heinreich's heart began to fibrillate causing him to lose consciousness at times. An advanced pacemaker helped keep his heart under control throughout most of the year, but the muscle tissue itself began to give out after Heinreich suffered a bout of endocarditis, an infection of the heart's lining, in late November. The bacteria had been unintentionally introduced during the pace maker's implantation procedure.

By January 2013 Heinreich had deteriorated so much he had no choice but to get a Mark-7 artificial heart put into his chest. The Mark-7 hearts were derivatives of the left ventricle assist devices and the Abiocor hearts of the early 2000's, and were approved by the FDA to serve as temporary heart replacements just before the end of the decade. The Mark-7 hearts were warranted for five years before requiring a replacement unit, and many patients, too old and feeble to receive heart transplants, simply held on to their Mark-7s until they died of some other cause, usually a fatal infection of some drug resistant bacteria.

After Heinreich's heart was replaced, the 98 year old could breath easy again. "You sound a lot better," Bob told him a few days after the implantation at the Texas Heart Hospital.

Heinreich, wired to all sorts of monitors and diagnostics, was on his bed in his private room reading a company report when his VP walked in. "I guess you can say that I am a heartless man Bob, or a tin man with a mechanical heart. Tell me, how are things going?"

Bob felt good about the news he had come to deliver to his boss. "Well, Charlie says we are a go for launching the first of the prototype collector satellites as scheduled for

February. Our Earth-based collector site in the Mojave is fully operational, minus of course, the microwave source from space. If everything goes well, and the collectors unfold according to plan, we should be generating 10 mega Watts of electricity by sometime in April. Charlie's back up plan should take care of any unfolding problems with the collector array should they occur. He subcontracted with Honda to develop a general robot that can be controlled telerobotically from Earth to help fix any unforeseen problems with the delicate solar collector array unfolding process."

Heinreich soaked in Bob's summary with a sense of satisfaction. He hated being bedridden—no matter how pretty his nurses were—and was pleased that he was getting back into the swing of things, even if that meant doing so using the wireless web from his spacious hospital room. "But what about the crazy environmentalists? Are they going to pose any problems to our launch?" Heinreich asked.

Heinreich was worried that some last minute lobbying might delay the launch of his long-awaited rocket. Several astronomical societies and environmental groups didn't want ALM's prototype solar collector put into space because it might block the view. What they didn't seem to understand was that the collectors were to be placed in geosynchronous orbit, and that even though they were large, they wouldn't block out more of the night skies than a distant planet would.

"Those idiots, driving SUVs I'm sure, want everything for free," Heinreich complained.

"They are not going to be a problem Heinreich," Bob broke in trying to reassure his boss. "Our lawyers are seeing to it. You'll see."

Bob was right. By September 2013, ALM had its first functioning solar collector and transmitter orbiting Earth and delivering power to the prototype station in the Mojave Desert. Given world conditions, the continuing violence in the Middle East of Islamic fundamentalism versus secularism, the killing of American soldiers still protecting US interests, the acts of terrorism, the plagues of AIDS and hunger ravaging Africa, parts of India and China, the escalating cost of petroleum based products, the increasing effects of Global Warming, not many people noticed the news release of the ALM milestone, especially after dozens of private, wealthy peoples had earned their astronaut's wings taking tourist rides into suborbital space—there was also the latest Hollywood sex scandal to distract people.

Not many people noticed ALM's other plans either. By 2017, ALM aimed at launching a small army of customized fifth generation Honda robots to assemble a small moon base to extract Helium-3 and build a fusion reactor. Physicists on Earth, working on the International Thermonuclear International Reactor (ITER) back in 2011, made key advances in demonstrating the feasibility of building commercial fusion reactors, and ALM was going to go full forward with Charlie's Helium-3 plans. There were only a few kinks left to get past in fusion physics, and these were thought to be mostly a matter that some careful engineering could resolve.

What people did notice, however, were variants of the customized ALM-Honda robots—and competitors—starting to do all manner of domestic chores, especially for the increasingly larger population of the elderly in economically advanced nations. Fuel cell powered robots—wirelessly connected to the internet—began working home security jobs, as well as dispensing medicines and monitoring vitals as transmitted by chips inside pacemakers and other implanted medical gadgets. If someone's heart failed, the robots would call 911 services and administer basic CPR, including shocking stopped hearts. For all this remarkable capability, however, the robots of 2013 still didn't have enough manual dexterity or artificial intelligence to do the hard things, like washing dishes, folding laundry, or holding interesting conversations.

***

The militaries of the world loved robots even more than gray-haired geezers pulling oxygen bottles around. Early generation, semi-autonomous spy robots went everywhere, under the sea, into the skies, and on the surface. They came in all sizes from insect sized to elephant sized. If these machines spotted something interesting, they could alert their masters— usually 18 to 22 year old soldiers—working as far as half way around the world in air-conditioned parlors. If these machines were attacked, they could defend themselves with non-lethal noise or tearing agents, or lethally with bullets, missiles and lasers—or torpedoes if they were underwater. Given the successes of the early models, the military forces of the United States, still fighting the war on terror, busied themselves building the world's first complete army, navy and air force of semi-autonomous robots designed to be controlled remotely by eager, young soldiers from afar raised on video games.

Thanks to the ever present war on terror, by 2013, Big Brother was everywhere public, and even in some places private. If a person went to a park, or the beach, or downtown to the library, or the mall, or the airport, or even to a bar, he or she could not doubt that he or she was being watched by computers running crime and suspicious activity subroutines designed to flag human supervisors of potential problems. At home, with computers now being embedded in every room and in nearly every gadget, all hooked together by wireless technology, a person could never be sure whether he or she was being spied upon by some remote hacker. Quite often a video web phone carelessly placed in someone's bedroom would be compromised by a hacker, only for the victims to realize after the fact that their sexual activities were plastered all over the web for a fee. Parents, fearful of child abduction crime, began to implant GPS chips into their children, and with Radio Frequency Identification Tags (RFIDs) pervading nearly everything, it became increasingly difficult to lose people and things.

On the medical front, most types of cancers were becoming far more curable. In the worst, rarer cases, tougher cancers were becoming more like chronic diseases than death sentences. For 50,000 dollars US, the wealthy were already getting used to the idea of having their genomes scanned, not only for current or future problems potentially written in their genes, but for the genomes of any harmful bacteria, viruses or cancers they might be harboring in their bodies. With the price of the process dropping 8% per year, it would not be long before most people could afford the testing, and the testing was getting speedier. From a month to get results when the technology first became available to the public, it was already down to a couple of weeks, and eventually, as scientists worked on the process, it was expected to get down to a matter of minutes before the end of the second decade.

Alzheimer's cocktails were beginning to show early promise, as were cocktails for almost every other neurodegenerative ill. After decades of work, there was even much talk in the medical field of finally mass producing FDA approved artificial electronic retinas to cure blindness. Others were pushing the first generation of so-called youth cocktails designed to strengthen muscles and encourage neural cell growth through the clinical trial systems of the world. These anti-aging groups were cashing in a decade's worth of biochip work comparing the genomes of peoples of all ages, and sorting out how gene activation networks changed with increasing age. Beyond the controversies of human cloning and stem cell medicine, the field of artificial life was bringing potentially new specters of doom and gloom to menace the world. Life had become something a researcher could create from scratch using only a computer, some basic molecular soup, and a general microfluidics processor.

The physicists weren't falling behind the biologists. Hans' grandson—Axle's son Paul Fritz—was finishing up his postdoctoral research at MIT. Paul, using a synchronization scheme that only he seemed to understand, was developing long-range teleportation technologies using quantum entanglement that far exceeded the limited teleportation technologies extant in his time. The governments of the world were already beginning to use quantum entanglement—a property of quantum mechanics that Einstein didn't like and dubbed as "spooky action at distance"—to perform unbreakable encryption tasks.

Paul—not caring which way the arcane debates fell with respect to the validity Bell's inequalities—wanted to do more than teleport secrets and a few quantum states of super cold atoms. He wanted to teleport things as large as DNA molecules using an improved version of the (Einstein, Podolsky and Rosen) EPR teleporter he had developed as a graduate student. That toy model, which had filled a small lab with all manner of cables, laser and computers, could only teleport the electronic quantum states of small molecules no larger than water across a room, but that was already a quantum leap that shook the world of physics. What he was missing to teleport larger things lay in notes his unknown great-grandfather

Albert Einstein had written back in the mid 1950s concerning Bose-Einstein statistics and EPR synchronization.

A trip to the MIT science library on a hot date with a pretty graduate student, Rebecca Peterson from the math department, lead him to a copy of those notes. They were hidden in a place well suited for making out, and when things got a little steamy between Paul and Rebecca, her long hair ensnared itself in the old binder spilling out the notes. "Rebecca, thy name 'tis not perfidy, but 'tis serendipity," he told his panting date with a grin. Nine months later, Paul's EPR synchronizing teleporter was born on paper, all the equations having been checked twice over. Fortunately for the world, Paul, smelling the gadget's potential, didn't publish his results. Instead, he founded a small company and sought out venture capitalists to help him develop his technology. He reasoned he needed about a hundred-million dollars in quantum computer and power plant equipment.

***

Macular degeneration began to rob Lise of her eyesight. By 2016, when she was 102, she was legally blind. She opted to have artificial retinas implanted in her eyes. More than just giving her a primitive eyesight capability—which wasn't much better than what you might have seen on a black and white monitor sporting only 200 by 200 pixels—her new retinas came linked to the global wireless world wide web. She could close her eyes, and anywhere in the world there was a web cam—atop a building, under the sea, or even in space orbiting the moon or on Mars—she could see a pixilated image of the scene being captured.

"I always wanted to travel when I was a young girl. Now I can tour the Solar System," she told Professor Pierre Lacombe a few days after her new retinas were activated.

Pierre, who was now 70, was a bit jealous. "I wish I could do that myself," he told Lise.

"I can't see why sighted people won't be able to do this in matter of a few years dear old friend. How is the youth cocktail making you feel? It has been a year since you have been on it hasn't it?"

Pierre showed Lise the muscle tone in his arms. "It's not bad Lise. Most of my muscle tone, as you can see has returned since I started taking the latest youth concoction. I feel as if I were fifty, maybe forty, and you, how do you feel on the new potion?"

Lise laughed. "Like ninety, maybe eighty." Lise was joking. Before she began her latest generation youth cocktail regiment, Lise was barely able to walk. Now she could stroll for several kilometers at a time, she could think clearer as well, and like Professor Lacombe, her muscle and skin tone were recouping some of their lost elasticity and strength. The only scary thing about the youth cocktails of 2017 was that no one really knew what effect they would have on human lifespan. Would people live to, say, 120 years and then drop dead, or would they liver longer than that? The administrators of the United States Social Security program wanted to know.

It was on 4 July 2017 that Lise stumbled into something that totally changed her life. Surfing the web with her retinal implants, she ran across hand written documents from the end of World War II that had been scanned into a historical archive at the University of Berlin using optical character recognition tools. Jews and other peoples who had been victims of the Nazis had, after the Allies won the war, written letters and posted them into a huge set of log books to help them find lost family members. When Lise began searching for Rebers, she eventually found a letter written from her father looking for her. Dr. Reber had submitted it before receiving the forged

Soviet notice of Lise's death—when he still thought she might be living.

Dear Lise, I don't know where to begin except to tell you—with every hope that these words will find you one day—that I am alive, looking for you and love you very much. I am so very sad to have to tell you that your mother and Olga did not survive. They were gassed Auschwitz. As for Heinreich, he too is alive.

Lise's jaw dropped.

I ran into him just yesterday at the Berlin rail station. He was captured by the Americans and is being sent to the United States to help them develop jets. He is looking good and is desperate to find you. Like you, he too is alone. His mother the baroness was killed by an American bomb, and his stepfather committed suicide after suffering a debilitating stroke. You can find me at my old clinic where I intend to continue working for a while. I hope to hear of you, Aunt Lilo and Uncle Elster soon. Your loving father, Max.

In her mind Lise screamed a NO that could have been heard around the world after she read her father's letter. To the horror of imagining Nanerl being shot while Belinda held her, now was added the horror of imagining Belinda and Olga being gassed in some dark chamber filled with screaming and gasping figures. Then there was the idea of her father living without knowing that she was alive. The Soviets lied to her about the deaths of Max and Heinreich, and, evidently, to Max and Heinreich about her death. The thought of Stalin in her mind made her feel like vomiting and killing at the same time before her thoughts turned to Heinreich, the love of her young life, and the life that he and she had been denied by Hans and by the Nazis. Even more rage filled the old lady's veins.

With tears streaming down her face, Lise frenetically began to pour through the web using any and every search engine to look for the paper trail that her father and her husband would have left behind throughout their lives. It didn't take her long to find her father's death certificate. Max had died in 1968 in Buenos Aires while searching for Nazi war criminals—Heinreich had taken his body back to Berlin for a burial beside Aunt Lilo.

The rush of emotions as she read these old documents was overwhelming, but Lise pressed on. Heinreich was still alive! His whole biography from the time that Captain Yeager had broken the sound barrier to ALM's latest details and FAQs was all on the web. Further research into newspaper articles that had been scanned into electronic formats revealed that Heinreich's life had been full of tragedy. He had married— the woman Donna looked so beautiful in her wedding announcement—and lost not only his second wife, but both of his sons, one as recently as 11 September 2001. Little wonder, she thought as she read through the ALM goals web page, that Heinreich wants to rid the world of the need for Arab oil. Then she searched for Hans, the little bastard that had destroyed so many innocent lives. He was dead, but his children and grandchildren were thriving, no doubt on assets stolen from so many liquidated Jews.

Lise, unable to support the weight of her emotions, stopped her surfing for a few hours. When she found Paul Fritz, the most evident member of Hans' progeny, she found his MIT doctoral research papers on EPR teleportation. Then she found Paul's EPR synchronization teleportation web page of his small company. "My god," she thought. "That technology, if it can be developed and made to work, it can be made to do very nasty things."

It wasn't long before Lise, overwhelmed with emotion, was drawn back to Heinreich. She poured herself a glass of sherry and laid herself to bed. She spent the rest of the afternoon looking at pictures of Heinreich posted on the web.

Holding the printed photos in her hand, leafing through them, she started crying not quiet knowing what else to do. When the sun dipped below the horizon in Paris, Lise didn't even turn on the lights, even when her robot tried to ask her if she was in need of assistance. Based on her heart rhythms, the machine sensed there was something not quite right with Lise. Only the next morning, as the sun arose, and Lise had spent the whole night reliving her memories, and contemplating so many, many possible lives she could have otherwise lived if only she had known the truth, she picked up the phone and called Professor Lacombe.

"Dear Pierre, you will never guess what happened to me yesterday," Lise began through sniffles. "I found my husband on the web, and he is alive. I found out so many other horrible things as well."

Pierre was stunned. He couldn't believe his ears as Lise related all her discoveries to him. Several times even he broke down in tears. "So what do you plan to do Lise?" he asked her.

Lise hadn't thought things out. "Perhaps let things be Pierre."

Pierre didn't like this idea at all. "You mean not approach your Heinreich? That would be so wrong Lise. Don't you think he has the right to know what happened to you rather than go to his grave thinking you never made it through the war?"

Deep inside Lise knew that her friend was right, but she was afraid and even felt guilty. "Shouldn't I have looked for these records when I was released from Russia?" she asked Pierre.

"Why should you have? You had their death certificates, and the other documents the Soviets gave you, and the world's old newspapers weren't scanned into electronic form and archived until years after your release."

***

Heinreich, motivated in great part because he knew his time on Earth was all but gone, scarcely allowed himself a private moment in 2017. Even though he himself was taking the latest youth cocktails, and was feeling much better, he knew that his second mechanical heart could give out at anytime. The first had nearly killed him when it seized during a routine diagnostic procedure at the Texas Heart Hospital. The diagnostic reader had short-circuited causing the CPU in the pace maker to fry.

Heinreich's senior vice-president Bob Levanthal retired that year at 76. "I want to spend more time with the kids and grand kids Heinreich," the gray-haired man told Heinreich. "You've have been a wonderful boss and a good friend."

It was a very sad day in Heinreich's life when Bob worked his last day. "Bob, I am going to miss you mein friend. Don't be shy and keep in touch." Other than Bob Levanthal, chief scientist Charlie Moss and Jana Menzel back in New York City, Heinreich had no one on Earth he could truly call friends. Jana, still running the trading firm in New York, now had three teenaged daughters of her own. Whenever he spoke to them on a video phone, the girls would call him grandfather. Heinreich liked that.

On 5 July 2017, the last thing Heinreich was expecting was a call from the past. Heinreich's robot rolled into his gym where its master was exercising on a treadmill, and informed him that he was receiving a video call from Paris, and that the sender's name was Reber, formerly of 17 North Strasbourg Street in Berlin.

"Is this a joke?" Heinreich asked. The machine, of course, didn't understand its master's question. Heinreich hadn't heard that address for nearly fifty years, ever since Dr. Reber had died. It was the Rebers' old address.

With his curiosity piqued, Heinreich stepped off the treadmill and walked into his office clutching a towel. The large window facing the hanging gardens had turned opaque and was flashing the words, Incoming Call. "Accept," Heinreich blurted out standing in front of the screen. What he saw transfixed him. It was a computer generated avatar image of Lise as she had appeared in 1939 the day of her departure to Paris. "Is this a joke?" he asked as he stared on numbly at the image.

"No. It is I, Lise von Onsager, as I appeared the day we were parted. I thought it would be easier for you this way dear Heinreich."

It was obvious to Heinreich that the computer generated image of young Lise was trying not to cry. "We were both lied to," Lise continued while Heinreich listened in near shock. "The death certificates you and my father received were false. I was sequestered by the Soviets to work on their nuclear weapons program. I was similarly given news of your death."

When Heinreich heard these words, he felt his legs turn wobbly, and sat down before the enormous image of young Lise. "Mein God," he mumbled as Lise explained her discoveries. Her words throughout the next few minutes were some of the most painful words Heinreich had ever been subjected to, especially when it came to Donna and his son.

"If you will pardon my having "spied" on you with the internet, I'm so sorry about the death of your second wife and your children dear Heinreich—"

Heinreich stopped Lise in the middle of her sentence. "No Lise. It is I who am so sorry for you. I was part of that Nazi machine all the while Hans lied to me. I would have killed the bastard and found a way to save Belinda, Olga and your father if I would have known better."

Lise's computer generated image began to cry. "I know you would have dear Heinreich."

Heinreich himself broke down in tears. "But as the war pressed on and mein mother was killed by the Allies, I became angry at the world. I could not stand doing nothing any more. I raged at having lost you. I raged at Nanerl's death, and at my inability to help Max and Belinda and Olga. I raged at Hans. I raged at having lost mein mother. I raged at the Americans for having killed her and mein father. I wanted to fight and to kill, and I did. I fought and killed young American boys, and thus, without thinking or wanting to, I helped Hitler."

Lise did not want absolution. "And I worked on making atomic and hydrogen bombs. We did what we had to do."

Heinreich peered into Lise's young eyes. "May I see you?" he asked.

Lise understood his meaning. A few seconds later the image of the avatar morphed into Lise's real appearance, that of a very old woman.

"You are beautiful," he told her. "If I send a jet for you, will you come to Texas? I have to stay near my cardiologist so that he can monitor my mechanical ticker."

Lise, suddenly feeling a flush of teenaged emotion, smiled. "Of course," she replied.

When Lise arrived at Heinreich's gigantic condominium near the Texas Medical Center, she didn't quite know what to expect. She had the limousine driver ring the doorbell while she waited behind the tall young man with some hesitation. When Heinreich opened the door she was surprised by how old and frail Heinreich looked in person. Without saying a word, the two of them embraced for a long while as tears streamed down their cheeks. "That must have been some trip to Paris," Heinreich blurted out while wiping his tears. "It only took you, what, seventy-seven years to get back to me."

Lise wiped her tears and let out a little laugh. "I've missed your arms Heinreich."

Heinreich embraced Lise again, kissing her on her cheek. "I'm never going to let you go to Paris alone again, unless, of course you have a life back there," Heinreich whispered into the old woman's ear while he kept his arms wrapped around her.

"No my sweetheart. There has never been anyone but you." Lise stroked Heinreich's gray hair. "My life has always been with you."

Lise and Heinreich spent the better part of the morning telling each other their lifelong stories. The robot brought them cool refreshments every hour or so.

"When I let Donna come into mein life, I thought you were dead. She was a beautiful lady full of love, and she brought me back to life. When I had Karl and Douglas, I was the happiest man in the world. After the accident—having our tires shot out—when only Douglas and I survived, I hung on to him like a drowning man hanging on to a lifesaver. He was mein life. Never in mein wildest dreams did I believe I would outlive him."

Lise never let go of Heinreich's hand while he told her the story of Donna and the boys. "I never believed you were dead dear Heinreich. After enough years passed, I accepted that I would never see you again, even my intellect accepted your death, but in my heart I never thought that you were dead. That's why I remained alone I suppose, even when this wonderful young man Joseph tried to love me. Now what are we going to do with whatever little time we have left?" Heinreich already had the answer to that question. "Marry me dear Lise. Be my wife again."

***

The first week Lise and Heinreich were together again, Heinreich gave her a tour of SolDynetics and Autonomous Lunar Mining in Galveston, Texas. Lise was impressed at the progress that had been made since the first ALM solar collector prototype test project. "My people are currently negotiating multi-year, multi-trillion dollar contracts with Japan, India and China to supply them 15% of their power needs within 10 to 15 years. The United States, unfortunately, is still addicted to oil and has no real interest in alternatives."

Lise was surprised by this. "I suspect the Americans will come around. They can't continue to subsidize the cost of oil for eternity."

"The lunar facilities are still very rudimentary," Charlie Moss told Lise when she and Heinreich stepped into his office. "Our small army of customized ALM-Honda robots has built a small Helium-3 mining and purification facility, but as fusion physicists are still working out some pesky issues, there is no fusion reactor to build as of yet. We are, however, planning to send the latest, most advanced batch of ALM-Honda robots in 2020 to begin to lay the foundation for a fusion reactor in case the physics kinks get worked out soon. This time, we strongly suspect they will."

Lise was impressed with the look of the latest generation ALM-Honda robots that Charlie showed her down in the basement. Whereas the old robots looked like machines she had seen in films during the 1970s, with box-like shapes and tubular appendages, the new ones looked more organic, more insect like.

"This generation of ALM robots," Heinreich added, "incorporate the lessons we've learned from the previous two generations. The previous machines had difficulties assembling our three conventional nuclear reactors on the moon because they couldn't crawl around in some of the curved tubing. This generation was designed using genetic algorithms specifically for that problem."

Lise was surprised that ALM had built three conventional fission reactors on the moon. "I thought you were trying to build Helium-3 fusion reactors up there, and don't you have plenty of solar power?"

Charlie let out a little laugh. "Our goal, certainly, is to build a fusion reactor, but we need energy to power our robots, heavy machinery, and a Helium-3 purification factory. Our solar collectors are far, far too small and too inefficient to provide the necessary power, especially for the purification factory, so we had no choice but to build three small fission reactors. Unfortunately there is no oil up there."

Lise laughed. She should have known better. It took a lot of energy to purify fuels for thermonuclear weapons, so it went that it also took a lot of energy to purify fuels for thermonuclear reactors.

Heinreich next gave Lise a tour of SolDynetics main campus located on the other side of NASA's Johnson Space Center. "I'm proud to say that SolDynetics is doing quite well. The company has recently won a contract with the Department of the Air Force and the Department of the Navy to monitor the emissions of the smart dust sensors they will periodically scatter across the globe's hotspots." It took Heinreich fully two hours to give Lise an introductory tour of the sprawling campus by the Gulf of Mexico.

Lise, often hanging on to Heinreich's arm was more than impressed, she was proud of his amazing accomplishments. "I remember the days back at the Castle Academy when you used to tell me you were going to build airplanes. And you would have me teach you mathematics to help you design them. Now look at what you are building. Nothing less than the future."

After their lunch break, Heinreich gave Lise a teleconference tour of his deceased son's trading firm in New York City. "Douglas' company is thriving under Jana's leadership," Heinreich told Lise just before Jana appeared on the large monitor in SolDynetics' main conference room. Lise's first impression of Jana was that she was an absolutely stunning and very bright woman.

"The higher the cost of crude oil and natural gas rises, the more profits our people make," Jana explained after the introductions were over. While Heinreich took a call from the Vice President of the United States in an adjacent office, Lise and Jana exchanged questions and dialogue for about thirty minutes. The two women got along very naturally.

When Heinreich returned to the conference room, he was pleased to see that things had gone well. "Congratulations on your new granddaughter Jana," Heinreich broke in. Jana quickly projected a picture of the infant back to Houston. "She's so beautiful," Lise remarked. "I am delighted to announce that we got the CIA X-sat program by the way. Vice President Johnson just let me know from Washington."

**

The day that Lise and Heinreich wed for the second time, they did so at a small ceremony catered at the base of the Grand Canyon by a helicopter tourist company. It was attended by Charlie, old Bob Levanthal and his wife, Jana, her husband and their daughters from New York.

The first thing Lise told her new husband, after they kissed, was that if he wanted to live to see ALM's goals go to fruition, he desperately needed to get rid of his mechanical heart and replace it with one grown from stem cells. "That mechanical heart of yours is keeping you alive my dearest, but it is not letting you live."

"With matters of the heart my sweetest, you are the boss," Heinreich, glowing with happiness, replied.

The nuptials and their guests settled before a round table for a luncheon served next to the flowing waters of Colorado River at the base of the enormous canyon, exactly at the location where Heinreich had made his boyhood promise of bringing Lise nearly a century into the past.

"Fortunately for Heinreich," Bob continued with a tangent to Lise's thought, "the mindless, moronic, fundamentalist religious doctrine of the second President Bush, which thought it more appropriate to pitch embryonic stem cells into garbage cans than to use them to advance medical science, didn't delay the fruits of stem cell science too much."

"Thankfully the English and others—realizing that a blob of embryonic stem cells numbering from a few tens to a few hundred cells did not constitute human life—picked up where the Americans dropped off," Bob's wife added. Elizabeth looked much younger than Heinreich remembered her last. The woman was seventy-two years old, but looked not a day beyond a healthy, even athletic forty years on her latest, second generation youth cocktail regiment.

"Those idiots," Charlie agreed, "simply couldn't do basic arithmetic. A few hundred primitive cells cannot compare to the average human brain alone with its one-hundred-billion brain cells, with each brain cell sharing between tens to tens of thousands of interconnections between other brain cells, connections that would take many weeks, if not many months to form in a viable fetus. By percent alone, a blob of 100 stem cells, a relatively large batch, equals, what, 0.0001 percent of an average human brain and has zero percent brain connections."

Jana summed it up much more succinctly, "they were killers in the name of God."

"No," Bob broke in joking, "these people wanted to make sure that God's will be done. They wanted to make sure that those people God had chosen to kill with cancers, with MS, and so on, children included, died damn it! They didn't want dumb scientists arrogantly usurping god and saving lives."

"Not until it was their ass, or one of their kids' asses on the line at least," Jana added.

"Hell," Charlie began anew, "it was estimated that only a few hundred thousand "baby-boomers" lost their lives due to the delays caused by the Americans, including, possibly, that Reeves actor."

"I heard," Heinreich interrupted trying to change the subject, "that some angry conspiracy theorists thought the American stem cell delays were a ploy by the American Social Security program to cut down on future payouts." His joke brought the house down.

"Well, fortunately for my Heinreich," Lise interrupted, "we are past this silliness. Let us enjoy this meal among friends and family in this magnificent setting."

"Here, here!" Jana added as 'round the table glasses of sparkling, chilled Champaign were lifted into the air.

***

Lise was right about the superiority of a real heart over a mechanical replacement. Within weeks of Heinreich having received his own, biological heart in early 2018, his color began to return to him, and the full power of the anti-aging cocktails seemed to hit him full force. By the end of six months, he was as fit as a healthy 60 year old. "I am ready to climb mountains!" he declared on Lise's 105th birthday later that year.

"You are an old fool," Lise told her husband, but Heinreich insisted, and proved his point later that evening.

"I told you, dear, that I was ready to climb mountains," Heinreich boasted to Lise after he had made love to her for the first time since 1939. "After getting me to change my mechanical heart for a real one, do you have any more bright ideas for me?" Heinreich asked his happy wife after kissing the tip of her nose.

With Heinreich laying beside her, wrapped in her arms, Lise began discussing something that had been bothering her since she had stumbled across Paul Fritz's work. "I have two ideas as a matter of fact. First, get a web interface for your brain."

"Okay," Heinreich interrupted kissing his wife.

"Second, hire Paul Fritz, Hans' grandson before someone else develops his ideas for dark purposes. I've been following his synchronization EPR teleportation work, and it scares me. I read his doctoral work, and it was impressive, but his postdoctoral work, which he never published except for a few limited notes, scares me even more. He claims to have found a way to teleport massive objects with some kind of synchronization technology added to standard EPR cryptographical teleportation. He has tried to start a small company to license his technology, but the world of physics is shunning his work and calling him a quack. No one, thus, has given him the venture capital to develop his idea. But, based on my readings and calculations, I think Paul may be on to something, even though I can't quite make the full theory work."

Heinreich was very surprised to be hearing of Hans' grandson, and didn't quite understand what his wife was getting at. "Could you repeat that to me in German?" Heinreich asked.

"Like I said my dearest, if what Paul claims to have achieved is true, he has found a way to entangle a massive object using a synchronization scheme I don't fully understand, and transfer it between standard EPR teleporters. In principle, he could have a load of Helium-3 entangled on a teleporter on the moon and then teleported to one on the Earth."

If there is one thing Heinreich knew about his wife, it was that she was generally right about things. "If you are serious Lise, then perhaps we should pay this Paul a visit. If his technology works, then ALM will be able to develop it. It should allow us to squash the Arabs out even sooner."

Lise was both pleased and relieved that her husband was interested in the work of the grandson of a man Heinreich despised with all his heart. "We don't need to tell Paul anything about what his grandfather did to us and our families. It wasn't the young man's fault, and from what I can tell he is a good person," Lise suggested. She had had Paul spied upon by a detective agency.

"I am okay with this," Heinreich told his wife. "What is important to me is to see the fruition of my dreams, and..." he paused.

"And?" Lise pursued.

"To climb mountains! Many, many mountains," Heinreich stated as he drew Lise closer to his lips.

"You devil!" Lise let out before her lips met her husbands lips.

***

Charlie was curious when Lise—coming to his windowless office—talked to him about Paul's work. "Of course I have heard of Fritz' work, but it has been discredited hasn't it?" Charlie asked offering Lise a chair.

"This is not true,' Lise told him sitting beside Charlie at his desk. "Paul has never been funded. That is different from being discredited. What people claim is that his synchronization scheme would be drowned out by perturbations from virtual particles seething in the vacuum." Lise, using Charlie's notepad computer, then went on to work out the equations she thought were driving Paul's ideas, and pointed out to Charlie where she was getting hung up.

Charlie was unable to resolve the problem past where Lise had gotten to, but he was now curious, and that meant that he too thought there was something to Paul's strange ideas. "I agree with you Lise. We should pay Paul a visit," Charlie concluded.

"I'll set up a meeting," Lise stated.

Charlie then laughed. "Do you know what just occurred to me about entangled quantum teleportation Lise?" Charlie asked her before she took her leave. "Banks and governments have been using cryptography based on EPR teleportation for years. Now a variant of this technology might be used to rob those very banks and governments as if by magic."

Lise smiled. "I know, I know," she replied. "And this would only be the beginning. It's far too dangerous a technology. We mustn't let it fall into the wrong hands. I will call you as soon as I get a meeting set up with Paul."

***

"As you no doubt suspect," Paul stated to Lise and Charlie at his tiny lab a week after Lise's discussion with Charlie, "my EPR work, on paper anyhow, has gone well beyond teleporting the electronic states of a small molecule. I have found a way to use entanglement to create exponentially growing synchronicity, and thus a way to teleport the electronic states of massive objects. Unfortunately, without money I've not been able to proceed with developing the scanners and teleporters up to the task, not to mention buy the necessary quantum supercomputers."

"That is why we are here," Lise told Paul.

"We wouldn't be here wasting our time," Charlie added, "if we didn't think you might be on to something. Lise and I have personally studied your dissertation and that portion of your postdoctoral work that you have made public."

"Very well," Paul continued as he fidgeted with his computer slide projector. He looked thin and sallow and acted resigned as he went through the motions of describing his work to Lise and Charlie. Paul had been trying to actualize his ideas for many years now, and, as was evident to Lise, he was extremely frustrated with the process of looking for venture capital.

Paul's presentation of his computer generated slides lasted about an hour. Lise, keeping mostly quiet, let Charlie do all the heavy questioning. She wanted to feel Paul out without revealing too much about herself.

"So are you here to help me develop my technology?" Paul bluntly asked at the end of his sales spiel, "or just to see how the grandson of Hans Fritz, former SS man and assassin, is doing Baroness von Onsager? Or should I have said Dr. von Onsager? I read about you in Professor Lacombe's book, and did my homework on who runs ALM and SolDynetics." Paul had, in fact, only learned about his grandfather's SS history in researching Lise, Heinreich and his companies after she had called him to set up their meeting at his lab.

Lise, rather than being taken aback by Paul's frankness, appreciated it. "I assure you young man, that my husband and I are here because we believe you may be on to something. We have no grudge against you. That would be plainly wrong. As to your grandfather, I would prefer never to hear his name spoken again. If you are interested, give us a call." Lise stood up and offered her hand to Paul before taking her leave with Charlie in tow. Charlie had wanted to stay and talk more, inspect some of Paul's equipment, but Lise, ushering him out, had not permitted it. "The young man needs to think things over," Lise told Charlie outside of Paul's lab.

"You mean, you want him to stew in his juices don't you?" Charlie stated more than asked. Lise didn't dignify Charlie's question beyond giving him a smirk and raised eyebrows.

It didn't take Paul too long to think things over. From a demographic point of view, he was definitely a young man relative to the large baby boomer population. From other points of view, he was quite old. Paul was 36 years old and still living alone in a tiny apartment, broke, and tired of eating TV dinners after teaching classes at a community college. There was no one special in his life. The last time he had had sex was with a student who had wanted her A- turned into an A+.

Paul also knew that the only way he could develop his ideas was with the proper funding, about a hundred-million dollars he estimated for equipment capable of teleporting massive things the size of bacteria. The only problem he had was that he felt awkward with the idea of working with people his grandfather had hurt so very badly. Were this Lise and this Heinreich out for retribution, he wondered. Would they steal his work and then leave him in the cold? On the surface, the woman Lise had seemed to be nice, but she definitely had treated him with reserve. The industrialist Heinreich was an unknown quantity.

After a week of stewing, Paul called Lise. "I'll agree to develop my technology for ALM on one condition. I want to meet with the Baron von Onsager. If I like him, then I will build you a machine."

The meeting between Paul and Heinreich went better than Lise had expected it would. The "kid" had something Heinreich wanted. Heinreich had something the "kid" needed. The two men understood this symbiosis upfront, and it was good enough to get things going.

***

The road to mindspace war was paved in many different, apparently unrelated, and seemingly innocuous ways. The years continued to pass from 2018 onwards, as science and technology advanced at an ever accelerating pace. Centenarians like Lise and Heinreich became the fastest growing segment of the population of advanced nations, while life spans in Africa and other poor regions continued to decline due to mutated forms of AIDS. An African man could expect to live to twenty-two before dying.

In advanced countries, life continued on the outside, at least, much as it had for tens of thousands of years. Lovers made new babies, some by less controversial partial cloning techniques. Regional wars and conflicts were fought. Catastrophes and pestilence, always the signs of the end of time, were as rampant as ever before—only now they were covered by CNN 24/7 and delivered wirelessly by PDAs, eyeglass monitors or smart phones to the joy of priests, ministers, pastors and mullahs worldwide. Repent was the word, for God and his new kingdom, as it had been for millennia, was just the around the corner.

How did the post Darwinian invasion that ended corporeal humanity begin? By a slow and insidious infusion of innocent augmentations Charles Darwin could hardly have foreseen, and

Heinreich, starting with his cochlear implants and later his artificial heart, and so on, like Lise and so many others with their increasingly advanced implants, had unwittingly been part of the invasion that ended corporeal humanity.

When the first of the Alzheimer nanochips entered clinical trials in 2027, they were hailed as miracles. The day was soon coming—the news services announced—when little mimic bio-silicate brain cells would live in people's brains, monitor individual neurons, learn to mimic their input/output patterns, and then replace them when the neurons died. For the first time in humanity's history, the brain/machine fusion barrier would be ended. The victims of early onset Alzheimer's would never know that their original brain cells, upon apoptosis, were being replaced by super-substitutes. In principle, after so many years, these people would end up with entirely new, high speed biosilicate brains guarding old connections, personalities, memories, passions and so forth that could be wirelessly linked to world's web systems and backed up on redundant, so-called mindspace servers.

As time continued to pass, bringing with it ever faster exponential change, Heinreich's and Lise's growing wealth approached 27 billion dollars, a large sum to be sure, but small relative to the trillionaires of Microsoft and of other such corporate giants. These ambitious trillionaires, after the oil embargo of 2023, suddenly started pouring money into Heinreich's vision of cheap energy, especially after the sinking of several luxury ocean liners and oil tankers by Islamic terrorists. The Japanese, Indians and Chinese were satisfying nearly a third of their energy requirements using alternative energy sources including tried and tested ALM products. These nations were well on their way to cutting ties with the Arabs and the increasingly unstable Middle East. The United States and many of her allies—though these nations had developed wind power and other alternatives to some degree—still suckled oil from Arab teats to maintain their livelihoods.

In 2029, the first of the commercially available, biologically inspired spintronic (electron spin, quantum computing) nano-scale brain augmenting neural net chips for Alzheimer's patients became generally available for public consumption on the black market. Despite government warnings that not enough was known about the long-term effects of the technology, it wasn't long before stock market analysts, doctors, scientists, and even artists—all perfectly healthy people—began to implant these newly developed, evolving wireless web-connected chips into their heads. People simply couldn't afford not to. The competition was murder. In principal, a good stock market analyst could price a complex, stochastic financial hedging instrument, requiring quadrillions of complex simulations, simultaneously cheat on his wife with a prostitute located halfway around the world—with his penis' nervous system being stimulated at the speed of light over the web as if he were actually enjoying the services of the prostitute—and hack into a competitor's mindspace system.

Heinreich received his own "memory" implant because at 116, he felt he was becoming forgetful despite his daily dose of ever newer generations of tweaked youth cocktails. Lise followed suit the next year. Suddenly Lise and Heinreich—stimulating the appropriate areas of their hybrid brains—could see and feel themselves as they were when then were a young couple. The first thing the old couple did was to transfer parallel copies of their minds to nearly synchronized, redundant mindspace servers. It took them some getting used to living various echoing realities in time-delayed parallel servers, but now, barring a planet-wide disaster, they became, effectively, beings as close to immortal as humanly possible.

With Heinreich's and Lise's brains and nervous systems wired directly to the spintronic world, which stretched from probes as far away as the frigid surface of Pluto to deep in the oceans, to anything anyone could post in virtual space, like a simulated trip into a black hole, they could experience things no "un-wired" humans could ever imagine. The first time Lise and Heinreich "walked" on the moon telerobotically via two ALM-Honda robots transmitting signals directly to their mindspace servers, they were surprised by how soft the sand felt on their virtual feet.

Creation and creativity thus exploded much in the same way as the creativity of the rock stars of the 1960s expanded under the influence of hallucinogenic drugs. Where one mind ended and another began became murky as humanity networked itself closer than it had ever done since the invention of the telegraph and trans-Atlantic cables. For Heinreich and Lise, this meant one last sad realization.

"Oh my God," Lise exclaimed as she peered into her husband's mind and memories for the first time. "The woman in the Soviet uniform that you saw in the rail station the day you ran into my father in Berlin was me! All of those years gone by with us torn apart, thinking we were dead, they needn't have happened that way if only you had pressed on. What on Earth was my father talking about when he told you it wasn't me?" Heinreich and Lise searched their memories.

"Perhaps," Heinreich suggested as his mindspace server reconstructed his old memories, "it was that woman standing beside you. If I put together your memory of the scene with mein own, and think about the direction from which your father approached me, he must have been talking about her. She did look like you from behind. I remember Max told me that he had tapped her shoulder only to be sourly disappointed.

She must have been the woman he thought I was seeing when I was actually seeing you just behind her."

"We, my poor father, you and I were so, so close my dearest," Lise lamented with a sense of deep bitterness and loss for what could have been.

A brave new world was thus being born. Everyone who could afford the technology thought it was great. Not many people worried about the possibility of a whole new kind of economic warfare—mindspace war. Instead, those who had transferred their minds to redundant mindspace servers scattered across the globe, were too busy celebrating the end of death. Life spans, barring a major catastrophe, were now to last into the indefinite future. A few of the avant guard even went so far as to ditch their physical bodies. Some of these lesser daring folk, however, only went so far as suspend their bodies in hibernaculum devices.

During these times, the power, and the true meaning of evolution, not in the spirit Darwin had imagined it, of gentle, accidental adaptation, but how Andrei Sakharov had once described it to Lise, of active, intelligent competition, reared its beastly head in all its glory. Darwinian evolution, starting with the single-celled beasties, was for the animals. It ended its long journey with Homo Sapiens Sapiens. Sakharov's flavor of evolution, ever accelerating, was for man by man and his sentient fusion with his computer systems.

BEYOND DARWIN—MINDSPACE WAR

In the year 2036, when about eighty percent of the peoples of the advanced economies of the world had their minds uploaded into internal mindspace servers, was the year the first Type I cyber murder was committed. No one who was on or near the planet missed the trial. Type II cyber murders—using the wireless web to, say, cause a pacemaker to stop a heart or change a critical prescription with fatal results—were old hat, the first being committed before 2017.

Type Ia murder, as hastily defined under the blind eyes of the global courts, which were caught off guard, was defined as the using of one's combined natural and bio-nano spintronic mindspace resources to infiltrate, overwhelm and destroy the mindspace of another person, thus rendering the victim nonexistent, or more simply put, dead. Type Ib murder was defined as the use of one's combined natural and bio-nano spintronic mindspace resources to infiltrate, overwhelm and ultimately control the will and mindspace of another person, or more simply put, to make the victim a zombie.

The first Type I murder was a murder of Type Ib. A Mr. Petifor Williams, a senior stock broker at the brokerage house of Stuton, Moore and Phillips, illegally appropriated a portion of his company's massive computer resources and secretly coupled them to his own personal mindspace resources. He then broke past relatively weak firewalls and took over the mindspaces of several dozen junior stock brokers who occupied only minimal mindspace server resources. Mr. Williams then forced the zombies under his control to execute highly speculative trades in the silicon markets thereby causing a severe collapse in the valuations of the technology markets, and made significant winnings when he cashed in his silicon futures. Mr. Williams succeeded by preserving just enough of the exterior personalities of his zombies to fool everyone for months—wives, girlfriends and husbands included—before he was caught after a freak power outage liberated the junior traders from their enslaved states. The liberated junior traders, feeling raped, went straight to the virtual police before Mr. Williams could reacquire his control over their mindspace servers.

The trial of Mr. Williams, which was broadcast planet wide, as well as to the Earth's three tiny space stations and the two small lunar outposts, quickly devolved into a terrific mess when one of the wives, Kate Stevens, of one of the victims filed a claim of rape against Williams. After filing her complaint, Kate Stevens, who was a very attractive woman, told the world during several televised tabloid interviews that, "every time I thought I was making love to my poor husband, it was the brute Petifor Williams who used me. I want to make him pay for each and every time that I was brutally violated in so many unmentionably wanton ways." Those "unmentionable" ways in which Mrs. Stevens was used appeared in a hastily produced, X-rated docudrama downloaded from her mindspace server that made her a millionaire by the end of the month and a staple on the web's talk show circuit.

Not long after Mrs. Stevens filed her complaint against Mr. Williams, Mr. Ricardo Sanchez, the husband of another victim, filed a complaint for estrangement against both Williams and the victim woman, Patricia Sanchez. Ricardo claimed that his wife Patricia had not been restored to her former self after Mr. Williams lost control of her mindspace server. "She is not my wife. To me and the children, Patricia is dead," the pot-bellied forty year old kept repeating on the television circuits. "This woman walking around and claiming to be Patricia is an abomination and scares my children. I therefore have a right to claim the wrongful death damages of my wife as well as to Patricia's life insurance."

"Poor woman," Lise told Heinreich after switching off her mindspace link to the news stream. "My good friend Andrei Sakharov once told me that we humans are cockroaches looking for any and every niche to explore and exploit."

"He was damned right," Heinreich agreed with disgust. "I wonder what will happen to the poor woman."

Mrs. Sanchez was forced into filing a counter suit against her husband for custody of her children. To her astonishment, she was told by the New York Supreme Court that before she could have rights to her children, she would have to demonstrate to the court by a preponderance of evidence from family members, friends and associates, under an extensive psychological examination process, that beyond a reasonable doubt, the victim's characteristics were restored wholly and with little or no significant corruption.

In the world opinion circuits, everyone felt sorry for Patricia and despised Ricardo. If Patricia could not be legally proven to be restored to her former self, she, unless her family or her friends took her in, would be left out in the cold, having no access to any of her prior possessions or privileges or even her own children.

More legal complaints of this nature broke out against Mr. Williams. Even the firm of Stuton, Moore and Phillips was itself drawn into the legal fray when angry silicon market investors sought damages against the firm for fraud and misrepresentation. And not before long, several world governments, interested in passing laws to prevent such dangerous activities, embroiled themselves in the mess Mr. Williams had started.

The exploding drama made for great, planet wide soap opera. It also made for planet wide paranoia. Could wives trust husbands and conversely? Could employees trust senior management and conversely? Who was real and who was a possessed zombie? And who was out there trying to make zombies?

"When you died in the old world, you were dead," Charlie remarked to Paul as they watched the trials unfold from their enormous teleportation laboratory. "Now, who knows?"

"Would you like to return to the old world?" Paul asked while trapping an error in his EPR synchronization quantum algorithm.

"No," Paul replied flatly. "But you can bet your ass that I will be getting my mindspace servers better firewalls. I've been slackin' on my security."

***

"Have you seen the latest news about Patricia Sanchez?" Lise asked her husband.

"No," Heinreich replied. "I've been preoccupied with our new lunar plans."

"She lost," Lise reported feeling a virtual pit in her stomach.

Lise and Heinreich were still living in Houston, Texas in 2036. They had long since moved out of Heinreich's condominium facility near the Texas Medical Center and into a huge complex sporting its own fuel cells to power things including Heinreich's and Lise's main mindspace server banks.

"It is very sad mein sweetheart, and it is most frightening what this murder of Type I has brought to the world. Are we, I wonder, on the verge of a third world war with every combatant literally being every uploaded human?"

Lise and Heinreich, though they still used their old bodies, had most of their growing mental faculties running on their mindspace servers, which they constantly upgraded with the latest technology. When they saw each other, they no longer saw their present physical selves. Instead, they preferred projecting images of themselves as they had looked when they had lived as newlyweds in Augsburg. Even the way they saw their home was a projection of their old Augsburg cottage replete with Lise's little flower bed. One key difference was that Hans no longer appeared in the photograph he had smashed 103 years into the dark past of Nazi Germany.

"I'm afraid dear husband, that the active Darwinian genie that Sakharov foresaw when I was in the Soviet Union has been let out of the bottle. If you have not noticed, all over the global wireless web, not unlike the days of PC security vulnerabilities and their attendant viruses being posted over the world wide web, are being spewed the methods of Type I murder. Everyone is downloading the various techniques. What do you think is going to happen my husband?"

Heinreich had only a few key words to say. "As I said I fear, I think we will have war for mindspace resources, the new measure of wealth."

***

Heinreich's fears proved themselves well founded. Within a few a few years of the Williams' trial, despite the best efforts of Earth's governing bodies to regulate the fuzzy, borderless universe of mindspace, the Earth erupted into an eat-or-be-eaten world. The forces of post-Darwinian evolution—based on active, intelligent, ever improving, ever accelerating science and technology, not on slowly working, old fashioned mutation and selection processes—were indeed out of the bottle. On Earth, it became a new dinosaur-like era all over again.

Wealthy peoples purchased as many huge, redundant mindspace servers as they could afford. Some of these wealthy peoples, like the carnivore Tyrannosaurus Rex, not satisfied with what they could afford to buy, went on the prowl for prey—the smaller, weaker mindspace reptiles. Often these carnivores would fight amongst themselves for mindspace hunting ground privileges.

Other wealthy peoples preferred a more herbivorous, school of fish model, herding together into huge groups that could fend off attackers. Often these networks had links that spanned the physical world, and the school of fish model, worked well for the most part, but it required continuous vigilance from within and without. One bad apple within the group who, like an insidious cancer, could be corrupted by an outside source, could bring an entire organization crashing down into ruins.

The middle classes also split into carnivore and herbivore camps exploiting the niches at their respective scale. The poor were, in a word, screwed. Even those peoples who hadn't had enough money to purchase the most rudimentary of entry level mindspace servers were caught in the fray. A street person's brain, once captured and properly conditioned, became a valuable commodity to add to someone's mindspace capacity. All one had to do to capture an un-uploaded human was to infiltrate his or her body with a network of bio-nano bots and link them to one's own mindspace.

New mafias erupted onto the scene offering "customers"

protection in exchange for half of their "customers'" mindspace resources. Many of those who refused these offers paid with their lives. All over the globe, people—their mindspace servers taken over or destroyed—ceased to exist by the thousands every second of every day as Malthusian anarchy ensued. In principle, no one on Earth was safe.

***

Heinreich, having just concluded a virtual meeting with his chief scientist Charlie and with Paul, took his wife in his virtual hand. "Paul and Charlie have come up with some important ideas for how we can get away from all of this madness on Earth—literally so. They propose, and I concur, that SolDynetics and ALM go into a space-based personal mindspace firewall business."

Lise started receiving Heinreich's projected thoughts as he went on.

"With Paul's superior EPR encryption and teleportation technologies, and our space capabilities stretching all the way to the moon, I firmly believe that we should go into the secure spaced-based mindspace business. Charlie, as you can see, has already put out some feelers and people all over the Earth are beating down the doors to get their behinds the hell off of the Earth."

"I like the idea," Lise replied, "but what is to prevent the same problems from erupting in space or on the moon?"

"In principle nothing mein dearest beyond policing ourselves with great vigilance and applying a simple, draconian law," Heinreich proclaimed. "All members will be given a fixed, finite mindspace domain. Violators trying to extend their mindspace domains will be terminated, and their resources will be redistributed to all members of our society in equal portions. As science and technology progress, all members will be allotted increased mindspace capacity in equal portions. No wealth gradients will be allowed to exist."

"It sounds awfully stifling," Lise commented. "It sounds rather like communism, and I know a thing or two about this. In the long run, it doesn't work."

"But it will in the short run," Heinreich countered. "It must, until we can find a better solution, and it must do so soon before humanity eats itself alive."

Lise agreed. "So what, specifically, are your plans?"

"We are going to the moon. You and I have already been scheduled for lift-off in a few weeks time. We and our senior SolDynetics and ALM people will run things from our lunar base. We may need to give up our bodies and go completely spintronic, but there are enough resources on the moon that that shouldn't become an issue for a while at least. I for one am not ready to dump my body. Are you ready to become a lunar citizen mein dear?"

Lise had but one exception to Heinreich's plan. "My sweetheart, it seems we are only going to save those who can afford to be safe. This is wrong."

Heinreich, kissing his wife with a great smile, was ready for this. "The richer mindspace entities, as they can afford more powerful firewalls, are much more likely not to be contaminated with mindspace viruses and Trojan horses. Accordingly, they will join us on the moon. But because I know your heart my sweet Lise, we will save other peoples, as many as we can, most especially the children of the lower classes by putting them into mindspace servers locked into low Earth orbit until we can make sure they are clean, and until we can build sufficiently large mindspace server farms on the moon. In this way, we will form an underground "railroad" not unlike the one black slaves used in America, to ferret people first into space-based servers, then to the moon. We will do so until there is no one left to save."

Lise felt very proud of her husband. "You are a good man Heinreich," she told him as she began going over the logistics of his plan. "You deal with the lunar facilities, and I will deal with the low Earth orbit facilities."

"As much as you would like to," Heinreich warned his wife as he closed the link with her mindspace, "we cannot save everyone upfront. We only have so much capacity. You will have to draw up a list of candidates from a very large pool of equally deserving souls." It was a rationalization Lise would end up despising. Not to be dismissed, Lise and Heinreich were not unaware of the two ironies of their plans. They were going to build, if only for a finite period, a form of communist society that, right off the bat, began with classist elements. The rich were going to the moon. The poor were to be put in low Earth orbit until they could be cleansed of potential impurities. It smelled too much of the showers in Auschwitz.

***

Within ten very rushed months, SolDynetics and ALM had moved a thousand billionaires to a mindspace server farm on the moon via heavy lift rockets. That only a thousand billionaires had had their own, personal mindspace servers physically transported to the moon via rocket had to do with the fact that SolDynetics only had ten heavy lift rockets in their inventory, and each of these rockets could only carry the weight of one-hundred of the latest generation mindspace servers with their attendant power supplies.

At the same time, about ten-thousand poorer people's mindspace contents had been teleported into two-hundred

Earth orbiting mindspace server clusters via ordinary EPR teleportation. These orbiting mindspace clusters—which were filled to capacity—were really hastily converted SolDynetics spy satellites housing large quantum computers. Personalities and memories in Earth-bound mindspace servers were scanned, entangled, and then teleported to equivalent EPR units in orbit in much the same way as banks had teleported credit card numbers and financial transactions before the eruption of the mindspace war.

Paul Fritz' prototype synchronization EPR teleportation method played a role as well. The ten-thousand people who made it into Earth orbit also had their DNA entangled via radio wave to the two-hundred mindspace servers. Once entangled, Paul's synchronization EPR technology teleported their picogram sized DNA samples into space, whereupon the samples were reconstructed into sealed sample holders inside small cryo-units. Many people were still reluctant to give up their bodies, and having those DNA samples available to them meant that they could reconstruct their corporeal forms at some later date. Of course, the genetic information could have been stored as ones and zeroes, but having a physical copy was important to their psychology.

Heinreich had no doubt that the mindspace beings who had been physically transported to the lunar facilities were their original selves. He wasn't sure, however, if the EPR reconstructed beings living aboard his satellites were copies who merely thought they were the originals, or whether they were truly the originals. Many of the people who had been teleported into the Earth orbiting mindspace servers had felt the same doubts before being teleported, but they had had no choice if they wanted a chance to live free. It was the ultimate take it or leave it situation.

Once the ten-thousand mindspace people were uploaded into Earth orbit, they linked themselves into a massive defense grid. Straightaway, the grid busied itself with destroying all other orbiting bodies. It was a natural move as these sophisticated spy satellites had been armed with small missiles and powerful lasers for self defense against anti-satellite weapons. The people in the Earth orbiting mindspace servers thus provided both forward looking observation posts, and a formidable defensive shield for the lunar base.

As an added measure of insurance, Heinreich saw to it that all his ALM launch facilities were destroyed. Included in this scorched earth tactic were all government and military launch facilities his company had access to.

***

Heinreich knew well that his ALM lunar base, powered by solar collectors and three small conventional nuclear reactors, was not designed to support more than a few dozen corporeal beings. "Until we finish building the mindspace server farm, it is going to be a crowded situation up here," Heinreich told Lise. "Many of our guests are going to have to store their bodies in hibernaculums and remain confined to the lunar lander mindspace servers." As Heinreich and Lise had done, of the thousand billionaires who had made it to the moon, about one-tenth of them had had their suspended physical bodies transported aboard hibernaculums in a smaller rocket.

"The situation is not much better for our people in Earth orbit. I finished my tally, and I am saddened—horrified actually—at how many of the children left loved ones, parents and siblings alike, behind. I know only too well how this feels. I find it unbearable to think of whole families drawing lots to see who would live and who would stay behind in the hell that

Earth has become." Lise paused. "I feel like a filthy Nazi," she declared with a bitter taste.

"The last thing you are Lise, is a Nazi. Believe me," Heinreich told his wife trying to comfort her virtual self in his arms. "Can you tell me what the ratio of children to adults is?"

Lise projected the information into Heinreich's mindspace. "About ninety percent are children. I feared that any higher the ratio, and the adults would not have been able to control things. Paul's linear programming model confirmed the ratio. Adults use more mindspace resources, but children need more control."

"What a mess," Heinreich concluded. "I can't stop thinking about Jana," he confessed. Heinreich had been unable to rescue Jana, her husband, her twin daughters or her youngest daughter. Their loss only added to his pain.

Dwelling in one of the epicenters of the mindspace wars, New York City, Jana and her loved ones had been assimilated a week prior to the first lunar launches. The family had been destroyed shortly after Jana's four year old granddaughter was approached by a virtual mindspace clone of her first dog, a dear pet which was killed in an auto accident. The clone simulation of her dog was a Trojan horse. Once the child let down the family firewall to let the dog into her mindspace, it was over for them in minutes. Yet more faces of dead people to remember was all Heinreich could think the day that he learned of the death of Jana and her family.

"It is not your fault my husband," Lise reminded him. "As much as you wanted to, you couldn't possibly have saved everyone with your extended firewalls."

Heinreich nodded. "Let's get to work on finishing the mindspace server farm. The sooner it is finished, the sooner we can start teleporting the cleansed children in Earth orbit to the moon, and the sooner we can collect a new batch of Earth survivors into our mindspace satellites."

***

The dinosaur-like war on the surface of the Earth, and even within the depths of the seas, raged on after the ALM refugees left Earth. Just as with the dinosaurs, for every measure—offensive or defensive—that some clever mindspace being invented, a countermeasure would soon arise.

Alliances formed on all scales, thrived for a while, then were torn asunder. The only difference between the old dinosaur wars and the new one was the ever accelerating pace of the time factor. Instead of evolutionary changes taking place over the course of tens of millions of years, upheavals now occurred over increasingly shorter intervals, weeks, sometimes only days.

Things took a profound change for the worse on Earth when the then CEO of Microsoft was blindsided by his chief attorney, Leo Lowmark Pierce. Before the hapless CEO realized that his mindspace was being disabled and taken over, it was too late.

After Leo took over his boss's massive mindspace resources, he took to calling himself Primus One, because in one fell swoop he became the largest mindspace entity on Earth. As with Hitler and Khan, at the cost of increasing global entropy, Primus One found himself at the head of a vast, well organized army of zombies and semi-autonomous robots employed to do one thing—lower Primus One's entropy by increasing his wealth, power and influence in mindspace resources.

Just before humans began to convert themselves into mindspace beings, the net worth of the defunct CEO had exceeded the combined wealth of the poorest two-billion people. Still, as a human being, the old man couldn't eat many more calories per day—without getting sick at least—than the world's poorest person. That changed when the old man converted himself and his loved ones into mindspace beings. In his new form, it took several nuclear reactors to not only to run the CEO's own personal mindspace servers, but to run all his sub-activities. When he and people of his class converted to mindspace beings, the gradient of the wealth distribution, from the level of abject poverty to wealth beyond the dreams of avarice, increased, essentially, to infinity.

To keep from falling behind the developments of other colossal mindspace beings, a vast portion of Primus One's mindspace dedicated itself to controlling laboratory bio-silico-robots as small as viruses, to gargantuan machines as large as giant dinosaurs. He employed the virus scale bio-silico- robots to develop and mass produce ever more advanced spintronic chips to continuously augment his growing brain power. Yet other robots that he controlled with his mindspace ran his nuclear reactors, while elsewhere worm-like robots mined for more uranium. Primus One also used semi-autonomous robots to guard his extensive physical assets.

It was a runaway process. The more Primus One developed his scientific and technological capabilities, the faster he could use his resources to develop even more powerful science and technology. The only thing that held Primus One in check were the attacks of the world's other giants trying to destroy him as he tried to destroy them. It was imperative to each giant to prevent any of the world's other giants from growing so big, so fast, that the leading giant might literally use up all of the Earth's resources just to sustain his or her own life faster than the Earth could replenish them. The very same thing had happened to tens of thousands of species in earlier days as humans consumed resources faster than the planet had been able to replenish them.

About the time that Primus One emerged, Islamic fundamentalist terrorists still busied themselves with trying to bring humanity back to cavemen times. For the most part these Islamic fanatics, still refusing to augment their brains for fear of being contaminated by other cultures and ideals, still crawling in their caves, removed themselves from the post-corporeal human equation. When with great luck they managed to destroy a large server farm housing many minds residing in several thousand mindspace servers, they doomed themselves.

Primus One saw to it that the world's entire surface, the surfaces of the oceans included, were coated with an advanced smart dust network far more pervasive than the RFID tags of the 2000's. With this smart dust, Primus One was able to track down and monitor every organic being larger than an amoeba. When the Islamic terrorists were located, Primus One unleashed an army of hybrid bio-silicate locust-like killing machines on the fanatics. With poetic justice, not one of them—their particular evolutionary path a total failure—survived their worst sadistic nightmares. They were chewed down to bits of cells and bone dust, crumbling into a bloody paste as they gasped their last breaths and scratched their eyes out. Not many mindspace beings missed these animals when they were gone.

***

"I sure as hell am not going to miss those filthy apes," Heinreich told Lise when he learned of the demise of the Islamic terrorists from the orbiting mindspace servers.

Lise, who had experienced the liquidation of humans on a more personal level than her husband had, didn't say a word. Instead she changed the subject. "Are you ready to give up your body? Even as genetically augmented as it is, it so vulnerable. Look at what happened to those terrorists. In my opinion, it is time for the thinking reed of Pascal to shed its corporeal form. I don't want to lose you, and you have left so much of what I love inside its feeble brain, like your deepest memories. Even though that brain is now made of bio-silico-spintronic nanochips, I'm afraid its time has come. I am giving up my body, I've decided."

Heinreich was taken aback. "I like my body," he began. "I feel quite attached to it," he continued after a bit. "With a tongue I can still taste. And with a heart I can still pine for times past. Without mein body, I fear I might lose contact with mein history, slowly at first perhaps, but inevitably so. For the younger people, with so much less history to their lives, I suppose that the loss of their bodies means far less."

Lise smiled. "I agree with your sentiments, and it is okay to be afraid, but life, running at the speed of light, is simply redefining itself with other realities and priorities my sweetheart."

Heinreich did not know quite how to feel about Lise's avant guard idea of giving up their bodies altogether,

***

It took about three months for things to settle down for the lunar refugees. During this period, everyone kept their bodies on ice and their conscious selves in their mindspace servers aboard the lunar landers. The landers were relatively safe, and their fuel cells could power the mindspace servers for years if necessary. The ALM-Honda robots, meanwhile, busied themselves with finishing the preparations for the underground mindspace server farm.

The server farm facility was to be built deep enough to protect the mindspace servers from essentially all deadly cosmic ray and solar storm radiation—the banes of delicate spintronic circuitry. Several glitches with the power distribution system made the refugees nervous two weeks before they were to be transferred from their rocket bodies to the server farm. But by the end of the penultimate week, power was being delivered to the server farm from two of the three small fission reactors. The robots then concentrated on finishing the railroad track for transporting the mindspace servers from the lunar landers to the subterranean server farm. When the time to be transferred came, the lunar refugees—the most advanced descendents of humanity—couldn't help noticing the irony of being transported as so many lumps of mass by such a backwards technology as a train on rails.

Once the initial hassles of moving and settling the thousand mindspace server beings into the lunar server farm were dealt with, several virtual mindspace meetings where held between Lise, Heinreich, Paul, Charlie and other senior SolDynetics and ALM scientists and engineers to prioritize what next needed to be done. Among the more prominent members of these initial committee meetings was Air Force Major General Jacob Joos, who had been in charge of managing all the SolDynetics spy satellites for the Air Force and CIA before the mindspace war broke out. For years General Joos had worked hand-in-hand with SolDynetics to build up the Air Force's and CIA's spy capabilities, and Heinreich trusted the man implicitly.

The first two council-like meetings dealt with making sure supplies were adequate to maintain the lunar mindspace facilities, and how to best get those supplies distributed using the ALM-Honda robots. The third meeting dealt with defense issues.

***

"My concern, as well as a concern of many others," Lise began at the third council meeting, "is that what is happening on Earth not be allowed to happen to us. Most of the people in our little colony like Heinreich's communist model wherein we are equals, police ourselves and exact the ultimate penalty for mindspace domain violations, but they want details. I've also heard concern about what is to stop SolDynetics and ALM people from taking things over. We built the machines and control the facilities."

Paul, who was an aficionado of English history suggested a system of chattel. "We form into groups, with the members of a group being held responsible for one other. If someone in a group commits an aggressive act, the other members must give up the mindspace capacity of the errant member to the rest of the population. This will encourage self-policing. If all group members cooperate to attack the mindspace of others, the rest of our larger population will dispatch them. To reduce the probability of having a group go rogue, we shuffle group memberships at random times."

Professor Pierre Lacombe, whom Lise had saved, liked Paul's ideas with one exception. "The shuffling of group memberships at random times is a good idea. Now, though it may reduce the motivation for self policing, the idea of a group yielding a portion of their mindspace to pay for the misdeeds of a bad member will lead to wealth gradients. We don't want this. So drop this part and keep Paul's random shuffling idea."

The rest of group seconded Pierre's motion to adjoin Paul's

shuffling idea to Heinreich's communist model until something better could be worked out.

"What shall we call ourselves in the interim?" General Joos asked.

"The council of six," Paul suggested. "And we can project ourselves about a round table in King Arthur style." The other council members found Paul's round table idea amusing, but in the end, projecting themselves onto a round table is what they did.

"We must discuss the situation on Earth and how it pertains to us," General Joos then stated. When he had everyone's attention, he began describing the situation on Earth. "For the moment the Earth beings are too preoccupied fighting among themselves to notice us. According to our satellites, islands are being taken by single mindspace winners. Hawaii was the latest. Millions of people no longer exist. Only one being remains. Might this happen to the whole planet eventually? And might the winner then look at us as a threat to be dealt with?"

Lise recalled Sakharov's having noted that the speed of light would hinder the size of a being. "Hawaii is small General Joos," she began. "Australia is far larger, and still divided. A lone mindspace being might not be able to control it because it won't be able to maintain coherence between its servers over such a large area given the finite speed of light. For the same reason, it can't react quickly enough to attacks on its periphery."

While General Joos digested Lise's thoughts, Paul added, "If what Lise is saying is true, then Earth might ultimately be ruled by a small number of beings, perhaps one for each continent, or one for each chunk of land the size of Hawaii. Alliances will likely form. It might be NATO versus the Warsaw Pact again, in a manner of speaking."

General Joos pushed his other concern back to the fore. "But what about us? Will the one winner, or the alliances of Earth see us as a threat? And if so, what will we be able to do about it out here on the moon?"

Charlie finally piped in. "Perhaps not much General Joos. The Earth has us overwhelmingly outgunned in natural resources and weapons."

"That's my point precisely," the general said.

Heinreich had been thinking about the threat of Earth for sometime. "We can move to Europa. It has a liquid ocean, is full of resources, and it would protect us from cosmic rays. We can engineer life forms that can live there with sufficiently large neural systems to support our mindspace consciousnesses. I've checked with our science staff, and they claim it shouldn't take too long to build such life forms using the local materials. The problem I see is that it would take us years to get there even with our fastest rockets."

Lise was unaware of Heinreich's plans. "Form and function are inseparable," Lise cut in. "As long as we cling to the bodies that evolution gave us, we remain human. But if we take new forms to survive in the oceans of Europa, and progress our evolution along these new lines, we will—as my husband recently pointed out to me—forget song and dance, and food and spirits, and what made us human. This is a question we, each of us refugees, has to consider carefully."

"As Lise knows, loosing my old body worries me," Heinreich interrupted, "but like General Joos, I don't think the moon is far enough away from the Earth, and that likely means we should leave for Europa to become fish-bodied mindspace beings."

"If so, I agree that we will forget what being human means," Peter broke in. "And the blue ball we were born on—as beautiful as it is—will eventually mean little to us." The council members paused to see the Earth rising against the cratered lunar horizon.

Charlie brought the digressing, esoteric discussion back into focus. "We are talking about leaving the moon to save our butts. Is Europa far enough from Earth to do so?"

General Joos was confused. "What do you mean Charlie? After building ourselves an army of robot assemblers and soldiers using the resources of the asteroid belt and Jupiter's other moons, can't we build up a hell-of-a defense grid? Then we can have a standoff between Earth and Europa."

This was not obvious to Charlie. "Jupiter is big, putting out more energy then it receives from the sun, and there are a lot of resources in that area of space, but nothing like the resources of the Earth. The Earth is much closer to the sun, and thus has access to much more energy than we would in Europa. And energy means progress. Earth's mindspace beings could develop their science, technology and military capabilities at a far faster pace then we would be able to do. In time, we would get our asses kicked."

No one liked hearing what Charlie had to say, but they couldn't deny his cold logic.

"Well, can't we move to Europa and do the best we can to suck the local resources, build ourselves a space fleet and leave the Solar System for another one?" Professor Lacombe asked. "There are hundreds of other solar systems out there."

Charlie quashed this idea as well. "This is not a good idea. First of all, we are still primitive. Let us not be swept away by post-corporeal hubris. We are still very much the children we used to be, only a bit faster. I myself only left my body five years ago. Where will we be in ten thousand years, if we launched ourselves into the interstellar void with our simple nuclear impulse technologies? We would certainly suffer the ravages of cosmic rays throughout the millennia before reaching the nearest stars. And we would be bereft of substantive resources to repair ourselves during the voyage. More importantly, we would not evolve at a pace competitive with Earth. Out in space, we would be both energy and resource hungry all of the time, and could die out there if—"

"—Surely," a frustrated General Joos broke in, "we would evolve nevertheless, albeit not so rapidly with our limited resources. We could spend the time advancing our pure mathematics, our theoretical sciences—"

"—But only so far," Charlie countered. "Again, without substantial resources we would be left to ponder theory, but it is experiment which drives and confines science. Earth would have the resources to experiment and decide between competing theories. We would not be able to do this. We would go on developing piles of competing theories and not be able to decide which ones to eliminate. The Earth alliances, or the grand Darwinian winner of Earth, would, through experimental advances, advance at a far greater pace than us crawling through the interstellar void. And it or they, with their empirically driven knowledge might then overtake us and swallow us up."

General Joos got visibly upset with Charlie. "But this is an absurdity! A thousand years hence we would be deep into interstellar space, and insignificant specks in the deepest of oceans. Could we be found? And to what end? Even if we are found, would we not be ignored as insignificant mites as this civilization passes us by on its outward journey should it ever embark on such a journey? What value could we be to them?"

Charlie finally closed the argument of fleeing the Solar

System. "Suppose all of us, the advanced ones of Earth and us, the technological amoebas fleeing the solar system—I mean all of us—are trapped in a closed universe so that ultimately all resources will be consumed and all life will die. Then the super Darwinian winner, or the alliances, will one day swallow us for the resource, small as we are, before our time. And even if the universe is open and infinitely exploitable, is it not likely that the alliances, or the super victor, would use up the resources of the solar system, and then expand to usurp the resources of the nearest stars long before we could get to them with our current slow-poke means? Either way, open or closed universe, we are screwed."

Only Paul suggested an alternative which Charlie couldn't exactly argue against. "Perhaps, if we go to Europa and develop ourselves enough before Earth overwhelms us—they are, after all, busy fighting a war—we might tunnel into a parallel universe if such universes exist. The work of my grandfather Hans suggests this possibility."

Everyone at the meeting snickered, except Paul who was serious.

"So what do we do?" General Joos asked.

"It seems to me," Heinreich said, "that we should make ready to go to Europa. We will have more alternatives at our disposal on that moon than here where we might get nuked. It's just a matter of time before someone on Earth builds rockets powerful enough to carry nukes to us." Everyone concurred with Heinreich's thought.

***

Paul and Lise supervised the Europa "fish" project. They had preliminary results after several weeks and wanted to share them with the full council, who were anxious to get a progress report.

"I think you will be exited by our work," Lise began as the council members gathered beneath another beautiful vista of the Earth rising against the gray lunar horizon. She projected several slides into their virtual minds. "Paul, our small team and I began with trillions of variables to sort out before we could begin any serious simulation work. Only when these had been reduced, did we then create quadrillions of DNA codes and propagate their development using Dr. Lewis' genetic algorithms under the conditions we theorize exist in the oceans of Europa from previous space probes. Most of the beasts died. Those which survived the harsh conditions the longest before dying under the simulations, we bred into improved beasts. These first generation survivors, bear in mind, lasted only minutes. However, after repeating this process over quintillions of simulations, Dr. Lewis finally bred our ideal beast. She believes it can live as long as a thousand years without much need of maintenance."

Paul, taking over, then projected another set of slides into the mindspace of the council members. "Here is what those animal mindspace shells look like."

A manta ray-like beast with exaggerated diaphanous wings formed in the mindspaces of the council members. The wings spanned several tens of meters. Along with an enormous central nervous system and a long tail, the beast had several eyes optimized for infrared light.

"As you can see," Paul continued, "it's a 'flying' beast of colossal proportions that will do well in what we think the Europan ocean is like, and it will be able to support our mindspaces in its ample nano-silicate central nervous system. Once we occupy these animal shells, we will be able to link ourselves to one another via ordinary long wave radio. By the same means, we will also be able to communicate and control all our machinery, be it in the sea, on a Jovian moon, or in some orbit."

Professor Lacombe was taken aback by the projected images. "I never thought I'd be a cross between a giant manta ray and a jelly fish."

Lise, facing her old friend, reassured him that he would still be able to see himself as he so chose within mindspace. "Perhaps like Gary Cooper," Professor Lacombe joked to everyone's amusement.

"Once we arrive in orbit about Europa," Paul continued, "we will populate the seas with these beasts by using landers to seed the oceans. These landers will core the ice and introduce the embryonic zygotes. When those big beasties are ready, we will then transfer our respective mindspaces to them via Paul's synchronization EPR teleportation method, Heinreich's esteemed opinion notwithstanding."

"I take it," Heinreich interrupted, "that while we await the completion of the gestation period, we will have to bide our time in orbit. During this time we should use our molecular assembler nanobots and the newer generation ALM-Honda robots to extract raw materials from Jupiter's other moons. We should then build a network of satellites to orbit Europa, Jupiter and some of its other moons to form part of our Jovian alert and defense system."

"I concur," General Joos seconded.

Lise paused things to add to Heinreich's suggestion. "As Paul mentioned, we will remain in control of our technology even when we transfer ourselves. There is no reason we would not be able to continue producing satellites and other infrastructure in the Jovian system by remote control."

"Given how few ALM-Honda robots we own," Charlie piped in, "and the small scale our nanobot molecular assemblers work on, it will take a while before the ALM-Honda robots and the nanobots make enough copies of themselves to become a real industrial force."

"This is correct Charlie," Paul replied. "And there is no reason, once we have sufficient industrial resources, why we wouldn't be able to build a more robust, non-biological server farm at the bottom of sea. If everything goes well, we should become the masters of the Jovian system within a relatively short span of time, say five to ten years."

"Now, as for getting ourselves to Europa," Charlie began, "we need to worry about two things." Charlie was the head of the transportation portion of the Europa project. "We need to get our robotic supply ships launched ahead of us so that they can create the required biolab facilities to, so-to-speak, seed the seas with the zygotes. That way, when we arrive, we should be able to transfer ourselves immediately. I need to know what the gestation period is."

"According to Dr. Lewis, she thinks it will take about three months to grow them from scratch."

Charlie paused to project several new images. "The next thing we need to worry about is getting ourselves off the moon. Given what Lise has just said, it follows that we need to take off about three months after our supply ships depart. I'm confident that the ALM-Honda robots should have finished building all thousand lunar lifter rockets in about two months. Given the moon's feeble gravity, these ships will be much smaller and cozier than our Earth rockets, and as everyone knows, they will be powered by nuclear impulse engines using fuel from our three nuclear reactors."

Pierre Lacombe was the final council member to speak. "What of our ten-thousand people orbiting around Earth?" For some reason, everyone's virtual eyes turned to Heinreich after Pierre's question.

"I don't know yet," Heinreich replied. "I am working on it. I promise you that we're not leaving them behind."

ALEX ALANIZ, PH.D.

JUPITER

E ight months after the council leaders of the moon base had met for the first time, they were treated to a cold shock. The refugee spy satellites detected that Primus One was building a large launching facility near the equator in South America, as well as an armada of hundreds of rockets, each stuffed with robotic fighter craft capable of reaching the moon. General Joos compiled an intelligence briefing and presented it to the full council.

"I don't get it," Paul exclaimed when he saw the slides first hand at the emergency meeting of the council of six. "Why invade us? Why not just nuke us as Heinreich has feared?"

The other council members sitting at the virtual round table could see that Paul was clearly in shock. General Joos, turning to face the young man supplied him the likeliest answer. "Firstly, whoever controls the high ground controls the war. Primus One wants to control the high ground, that being the space near Earth as well as the moon. Secondly, I'm sure that he would like to capture the infrastructure that ALM has built here over the decades. Starting from scratch comes with a pricey learning curve. If he can destroy us, but keep most of the lunar assets intact, he can finish the original ALM goal to produce unlimited fusion power. Primus One would then be unstoppable."

"I only hope," Heinreich broke in clasping his wife's virtual hand, "that we can finish building our escape fleet soon enough. And we will need to build as many fighters and fighter pilot clone simulators as we can. Luckily I know a thing or two about fighters."

General Joos did not agree with Heinreich. "Building fighters will waste our resources, and from the looks of the satellite images, we don't have much time before Primus One launches his fleet, perhaps five or six months."

Heinreich defended his opinion. "General Joos, you were born in 1984. I was born in 1913 and fought against hordes of unprotected American bombers in the second world war. Without fighter cover, we cut those boys down to pieces."

"But," General Joos broke in, "our intelligence indicates our escape vehicles will be faster than their small attack ships."

"General," Heinreich countered, "we don't know that for certain, and by carrying our huge mindspace servers and supplies to Europa, we will be as those American bombers, big, fat, lumbering and unprotected. Primus One has to know that we are watching him, and he has to think we are going to flee. Though I'm sure he would rather eliminate the threat we pose to him now, here on the moon, I'm just as certain he doesn't want to risk us escaping only to threaten him anew at a later date. He will come after us with fighters of his own. I promise you that."

General Joos kept disagreeing. "The fighter force Primus One is building, according to our intelligence reports, is based on a dual role force. His fighters will be able to fight in the Earth's atmosphere as well as in space. They are designed for dealing with local, atmospheric and low Earth orbit assets. Those fighters, I must insist, are not for us on the moon. We are wasting time and resources building a fighter force."

Professor Lacombe interrupted the heated debate between

Heinreich and General Joos. "Gentlemen, our orbiting refugees—the children—are in danger are they not?" he asked.

"Yes," General Joos admitted flatly. "But they have some capability to defend themselves."

"Can you not teleport them to the moon Paul?" Professor Lacombe pursued.

"Not at this range I'm afraid," Paul replied feeling low.

"This is not acceptable," Lise then added.

"If," Charlie cut in, "we can get a ship nearer to them, then we can upload them into a rescue server, but we have no server capable of holding ten-thousand mindspace souls. Maybe we can manage a few hundred. This had been our plan all along once we completed a large enough server farm, but we've been caught with our pants down. We've done a lot, but we have so much still to do and we are out of time."

"Why can't we just tow them?" Lise asked. The jaws of all the men around Lise just dropped. "The satellites themselves have sufficient mindspace capacity. Since they are solar powered, we would have to supply power from our own generators. Charlie, shouldn't it be a matter of fairly trivial orbital mechanics to do this?"

Charlie, working through the engineering and physics constraints, quickly calculated a proposal. "We can use four of the rockets that got us to the moon to grapple fifty satellites apiece. The tricky part will be—but I think we can manage it—to prevent runaway oscillations from building up along the towline."

"Get to work on it Charlie," Lise finished with a broad smile.

"Any ideas as to how we man a fighter force Heinreich?" Professor Lacombe then asked. "Not many of us are combat pilots by training." It was a foregone conclusion that between Joos and Heinreich, Heinreich was going to get his way and build fighters.

Heinreich, grinning from ear to ear, was ready for this question. "One of the best pilots I ever met and flew with—and I'm not talking about myself—was General Burt "Bone Crusher" Pratt. Years back his wife happened to give me an old book of his after he died. In that book of his I found a strand of his hair and several skin cells. We are going to clone as many Pratt mindspace simulators from the man's DNA as we can to pilot as many fighters as we can build. I will train these simulators with all of my memories of how Pratt flew against me in so many mock battles throughout the years. I will also train the simulators with all the lessons I learned flying in WW II. General Joos will add his experience flying A-10 warthogs in the second war against Iraq."

General Joos smiled. "I don't object to this part."

***

It was not long after Primus One emerged on the scene that he turned his attention to the ALM refugees orbiting the Earth in spy satellites and building a lunar colony. He recognized them as a threat from the start. The two-hundred spy satellites had, after all, destroyed all low Earth orbit assets with their laser cannons. Thus the refugees could see what he, and everyone else, was up to, but he could not see them except as fuzzy dots through adaptive optics telescopes.

Unhindered as they were by the mindspace war, Primus One realized the lunar refugees were free to develop powerful weapons that could one day rain fire upon the Earth. But as always, Primus One's first priority was to the mindspace war raging on Earth. He had to attack and defend, and this occupied the lion's share of his resources, but when resources permitted, he worked on his plan to deal with the ALM refugees. Having no space facilities other than two utterly wrecked launch facilities in his possession, he had to start from scratch and did his best to race against time. If he could pull it off, he would take the moon and its facilities for his own.

Primus One began, like all good warriors, by learning as much about his enemy as he could. He scoured the Earth for information on each and every one of the refugees. He was pleased that the vast majority of them were mere children. The adults aboard the spy satellites were mostly there to keep the children in order. Only very few of the adults were military people put up there to run the spy sensors and weapons systems. The real threat came from the lunar refugees. The lunar refugees were the controllers.

Naturally, Primus One concentrated on finding the weaknesses of Heinreich, the founder and owner of ALM and SolDynetics, Lise, his wife the nuclear weapons designer and General Joos, the Air Force major general who had controlled all the Air Force's and CIA's spy satellite resources. He missed Paul Fritz, great-grandson of Albert Einstein.

Primus One hit the jackpot with Heinreich. The name of the jackpot was Peter Wohlthat. The man was dead, but DNA in his grave was intact, as were his extensive KGB records. Peter, Primus One learned, had been Heinreich's faithful wingman during the second world war. For ten years Peter had rotted in a gulag while Heinreich had thrived. Peter had tried to assassinate Heinreich, but, best of all, had only managed to kill Heinreich's second wife, Donna and his first son Karl.

"Design and build me a fleet of space capable fighters," Primus One ordered his army of semi-autonomous constructor robots. "Dub them Wohlthats. Use Wohlthat's DNA to build mindspace clone pilots. When they are complete, I will train them to my liking."

***

There is never enough time when preparing for an invasion. The day Primus One launched his fleet of robot lunar landers and fifteen-hundred Wohlthat fighters came all too soon for the moon dwellers and the spy satellite refugees. Only the robotic supply ships—designed to deploy and build a small undersea incubator facility for the manta ray mindspace zygotes—had been launched to Europa from the ALM lunar facilities, and that had only been two months prior. In another month the lunar refugees would have landed their rescue mission to recoup the children-laden spy satellites and started their journey to Europa.

"Do you think someone told Primus One what our schedule was?" General Joos asked the council of six.

"I hope not," Paul replied. "That would mean that we have a spy amongst us, which I don't believe is true. But there is no denying that we have been caught with our pants down. We're not supposed to launch ourselves for another month. Now we have two days before the invasion force arrives."

Heinreich paused to consider the whole situation. "How many of our mindspace server peoples have been loaded onto their rockets?" he asked.

"Only about a fourth," Charlie replied.

"Damn it! That means that Primus One's landers will be on the moon before we can possibly get all our people off." Heinreich was steaming hot. "It took the damned ALM-Hondas two weeks last time to unload all thousand of us onto the railroad tracks and the server farm."

"What about the children in Earth orbit," Lise asked with desperation in her voice.

"They've been left alone for the moment," General Joos told Lise. "I figure the game plan of the Earth force is destroy us first, then deal with the spy satellites at their leisure."

Lise looked at Heinreich. He read his wife's pleading expression without difficulty. "Sweet Lise, we have to deal with the defense of the moon first. If we die, then the children will die as well." Lise understood that for the moment no Pratt fighters would be dispatched to defend the orbiting refugees.

Professor Lacombe then asked, "is there any way that we can we hold the Earth fighters off? There are so many of them."

General Joos supplied his answer abruptly. "Impossible. I can't see how Heinreich and his meager collection of two-hundred Pratt fighters can fight off fifteen-hundred fighters."

"Don't be so glum general," Lise retorted as she took Heinreich's hand in mindspace. "My Heinreich had over a hundred kills in his day, and he says the Pratt clone simulators—which you and Heinreich are part of—are far, far better than he ever was."

"Whatever we do, we need to hurry up and hope for the best, as our choices have been decided for us," Charlie added.

When the attack fleet passed the halfway point to the moon, General Joos ordered the moon's long range 500 defensive missiles be launched in a simultaneous barrage. Twenty of these missiles were nuclear tipped. The other 480 missiles were supposed to act like shepherd dogs and coral the Earth fighters into clusters for the 20 nuclear-tipped missiles to destroy. Long range sensors aboard the Earth fighters, however, detected the missiles armed with thermonuclear warheads. Before the shepherd missiles could do their job, the fifteen-hundred Wohlthat fighters spread themselves out over a vaster distance than the 480 shepherd missiles could possibly cover.

Twenty of the Earth fighters sacrificed themselves by crashing into the moon's nuclear-tipped missiles. The remaining Earth fighters regrouped and pressed on their attack like a cloud of angry hornets.

"What's the good news?" Heinreich asked as he had his mindspace server loaded into a Pratt fighter by an ALM-Honda robot. Heinreich had dubbed the lunar-based fighters by Pratt numbers 1 thru 200.

"Most of the Earth fighters made it through. Only twenty of the bastards are gone, having destroyed our nuclear missiles." General Joos replied.

Heinreich was disgusted. "You know the Pratts and I are not going to be able to stop the landers from landing. How many of our thousand mindspace people are loaded onto their rockets now?"

General Joos checked the launching bays. They were abuzz with ALM robots scurrying about trying to load the rockets with mindspace servers. "About half now. The ALM-Honda robots are working as fast as they can."

"That's not good," Charlie piped in. "Only ten hours remained before the arrival of the Primus One landers."

Before Heinreich launched himself and his Pratt fighters into space he gave his wife a virtual embrace. "I love you mein dearest Lise. Remember that always."

Lise kissed Heinreich in virtual space. "And I love you my husband," she told him as he blasted off from the moon in a cloud of dust. Somewhere ahead of his nose, hidden in the brightness of the Earth's reflection, fifteen-hundred fighters were bearing down on his small fighter force.

"Let's go get them boys!" Heinreich exclaimed a few seconds after his dusty blast off.

"Yes sir!" the two-hundred eager Pratt fighter clone simulators shouted back in near perfect unison.

A few hours later Heinreich and his Pratt fighters got into a hairball of laser and particle beam fire with the leading edge of the Earth fighters. "I'm going to kill you this time you son-of-a-bitch," the lead Earth fighter blurted out over the radio. "You left me to rot in a gulag for ten damned years. The last time my bullets only got your whore wife Donna and your idiot boy Karl killed. It's your turn to die now."

Heinreich, recognizing Peter's voice, was, as Primus One had hoped, taken aback for several seconds as he sorted out his memories. "It was you who killed mein family?"

The instant Heinreich finished asking his question, the lead Wohlthat fighter broke out into a sinister, twisted laugh. Its laughter was then followed by the Wohlthat beaming old KGB surveillance pictures of the von Onsager household the day Peter had killed Karl and Donna. Forensic police photos of their bloody, mangled bodies followed a few seconds later. Lise and everyone else on the moon also received these signals as Heinreich flew on transfixed and dumbfounded.

"Yes. I killed them," the lead Wohlthat replied.

Heinreich's blood boiled. "I thought you were dead Peter, but now you're a dead man for certain," Heinreich shot back.

"Which one of us is dead?" one of the Earth fighters called out in a mocking voice. "We are all Wohlthat fighters!" the fifteen-hundred fighters burst out.

Only a string of explosions near his left side snapped Heinreich out of his shocked state. "Stay focused my husband," Lise radioed back to Heinreich, but only silence met her in reply.

"Pratt 2 and Pratt 7, form up on me," Heinreich ordered. "Let's go for the group of six guarding the pair of lunar lander rockets." He fought hard not to let his rage get the better of his decision making.

Laser, missile and particle beam fire poured out of Heinreich's fighter and from his escort Pratts. The Wohlthat fighters returned fire in kind. When Heinreich grazed one of the Wohlthats, he noticed they looked much more like ordinary fighters than what the spy satellite pictures had indicated. This was good. It meant they were a more compromised design than Heinreich had thought they were. The Pratt fighters that Heinreich and the clone simulators were piloting, on the other hand, were definitely not pleasing to the eye. There was no streamlining to them whatsoever. They were shaped more like cubes with protuberances sticking out from every end, but they were optimized to fight in space, and change direction on a dime.

The first six fighters which blew up into balls of blue flame were Wohlthats. When the first Pratt fighter took a mortal hit, the Pratt clone simulator managed to ram his fighter into a Primus One lunar lander, making for a spectacular flash of sparks and light.

All in all, Heinreich and his Pratts were able to completely disorganize the leading edge of Wohlthats within a few minutes, but the waves of Wohlthats behind the leading edge were able to change course and completely by-pass the Pratts.

"We're holding about a two-hundred Earth fighters back," Heinreich reported to the moon. 'We've destroyed twenty-four of them, but have lost three of our own, and there is no way we can catch the thousand that slipped by us. Get the ground based lasers ready. And get the Alpha plan readied. I figure you have about an hour left before Wohlthats and landers get there. Please make certain Lise is off the moon before that happens."

As Heinreich and his remaining Pratts fought their way back to the moon, destroying another thirty Wohlthats, the first wave of Primus One lander rockets began to slow down and enter their final approach trajectories. These robot ships chose their landing points using radar to pick out important lunar infrastructure locations.

"When we get over our launching pad," Heinreich announced to his Pratts, "we must do our best to defend the remaining mindspace rockets as they blast off, or they will get cut to pieces."

On the moon it was like Pearl Harbor all over again when the Wohlthat and Primus landers arrived, except this time the attackers did their best not to destroy any infrastructure other than the lunar defenses. Most of the forty lunar laser batteries were flattened without having had much effect on the attackers.

When the Wohlthat raiders found the ALM launching facilities, they caught about two-hundred lunar mindspace inhabitants still trying to blast off. Heinreich and his fighters, cut off by a thousand Wohlthats, were prevented from helping those trapped on the launching pads. Within minutes, the airwaves filled with shouts of panic as the launching facilities came under attack. "I'm hit!" one mindspace entity screamed out just before his heavy lift rocket exploded into a huge ball of flame and light. It was a horrifying sight for Heinreich to witness from above while he and his band of Pratts desperately tried to fight their way down to the surface.

Only another twenty or so heavy-lift mindspace rockets managed to get off the lunar surface before several Primus One lander robots began to occupy the pads. These twenty-odd unfortunates were mowed down before they could get very high. The carcasses of their remains rained down about five kilometers downstream from the launching site. Most of the remaining empty heavy-lift mindspace rockets were cut to pieces on the lunar surface, trapping their would-be occupants in the mindspace server farm.

***

The Primus One lunar landers were ugly machines. They were tall and ungainly, and packed three tentacle arms as well as missile, cannon and laser cannon batteries. As soon as they would touch down on the surface of the moon—kicking up dust—they would jettison their rocket engine and begin attacking the remaining defenses of the stranded mindspace refugees.

The hundred-fifty or so mindspace survivors that were left stranded in the subterranean server farm began putting up a valiant fight. They linked their mindspaces with the ALM-Honda robots and engaged the enemy forces in pitch battles all over the ALM facilities.

The first generation ALM-Honda machines were hardened, hydraulic brutes built to assemble nuclear reactors. They would stampede from place to place flattening the flimsier Primus One robots. With their hydraulic pincers they would also hack and rip into the bodies of any Primus One lander they caught, ripping the offenders to shreds. The second generation ALM-Honda models, built in 2022, were less hardy, but were faster an far more agile. These models would work in teams to snare the tentacle arms of the Primus One robots, rip these limbs off, then topple the attackers on their side. A half dozen custom ALM robots designed to do deep mine coring packed extremely powerful lasers. Direct hits by one of these robots—after the attendant fireworks—would leave behind smoldering lumps of molten metal. The first round of Primus One landers fared very poorly.

The next round of Primus One landers changed their

tactics and called in for "air" support. When the Wohlthats showed up, each time the ALM-Hondas tried moving onto open ground to go somewhere, they were mowed down from above. With the aid of Wohlthats, the Primus One lander robots quickly turned the tide against the ALM defenders.

***

"We've got eight-hundred souls aloft Heinreich, and Lise is safe," General Joos called out.

A thousand kilometers above the moon, General Joos and the remainder of the escapees, using their long range sensors, were treated to a view of hell on the moon. They saw the flashes of explosions, the rising dust columns, the crushed buildings, and robot corpses with a feeling of awe and incredulity. What had been their home now laid in ruins.

When half of the Wohlthats broke off from the lunar surface and started climbing towards the escapee fleet, General Joos ordered the fleet to run their nuclear impulse engines at full thrust. "Heinreich, break off and come escort us before we get chewed to bits," the general ordered. He was damned glad Heinreich had stuck to his guns about building fighters.

The Pratts, not carrying around the heavy wings of the dual role Wohlthats, were lighter and faster. They broke off from the surface of the moon and left the climbing Wohlthat attack force in the dust. Seeing the Pratts abandoning them was a blow to the lunar ALM refugees.

"How are our people on the moon doing?" Heinreich asked. He had lost contact with the ALM mindspace refugees.

"The last I heard from Charlie is that they are trying to do their best to buy our escape fleet the time we need to get away." It was a sad thought for Heinreich. He had gotten to know all the 1,000 lunar inhabitants during his time with them, but he cleared his mind and lead his Pratts towards the survivors.

"General Joos, I'm executing Alpha. I'm disregarding your orders and I'm going to take the Pratts to the moon and see what we can do to save our trapped people. You and the fleet should have no problems keeping ahead of the Wohlthats pursuing you. They're damned slow."

General Joos immediately fired back, "you can't do that Heinreich! You'll leave us naked without a fighter escort. You yourself said that to me. I order you to rejoin us and cover our retreat."

Heinreich ignored the general's order as he directed his 130 remaining Pratts back to the moon. As he changed course, Lise came over the radio. "Heinreich, you must rejoin us," but Heinreich ignored his wife as well. Plan Alpha, along with its attendant feint, was on.

As the Pratt fighters closed to within two kilometers of the lunar surface, with a cloud of Wohlthats charging behind them, Heinreich yelled out, "Execute Alpha!" Suddenly all 130 Pratts decelerated as hard as they could and reversed directions. The much more massive Wohlthats, still descending, flew right past the Pratts heading in the opposite direction. When the Wohlthats finally managed to turn themselves around, they were only about half of a kilometer over the lunar surface.

"Goodbye Charlie. Goodbye mein friends," Heinreich called out into the void. The mindspace survivors on the moon, whittled down to fifty-four, who could receive Heinreich's voice, but not transmit, all knew what Heinreich meant. A second later a blast wave of neutrons and gamma radiation ripped Heinreich's main lunar facility to smithereens, ripping a hole into the moon two-hundred meters deep. The hydrogen bomb's explosion was powerful beyond imagination, and destroyed over three-hundred Wohlthat's directly. Another ninety lost their control systems when the electromagnetic pulse (EMP) fried their guidance computers. These crippled Wohlthats either crashed onto the moon or drifted into space.

***

The EMP hardened Pratts and mindspace rockets—built to withstand cosmic rays, solar storms, and electromagnetic impulses for the long journey to Europa—remained unscathed.

"I had hoped we wouldn't have had to use Alpha with real people on the moon. For all the times we ran this scenario in our simulations, I was not prepared," Heinreich told Lise and General Joos over the radio.

"I know my Heinreich," Lise replied. "It would have been better if we had all managed to make it."

"But everyone of us agreed that if any of us didn't make it," General Joos reminded Heinreich, "we'd buy the others time before destroying ourselves."

Knowing this full well didn't make it any easier to accept the losses for Heinreich. "How many of our nuclear bombs did we manage to get off the ground?" he asked.

"None my sweetheart. Most of them were destroyed before they could be loaded in the last cargo rocket. The rest fell into enemy hands."

We're screwed, Heinreich thought. He didn't say it though. After a few moments of silence, Heinreich called out to the nearly eight-hundred remaining survivors. "I'll be with you soon mein friends, as will our remaining Pratts."

When Heinreich caught up to his mindspace heavy lifter, he gladly uploaded his mind into its larger spintronic computer. The mindspace server aboard his Pratt fighter, though it could fit that part of his brain that had been human, couldn't fit much more, and it made Heinreich feel claustrophobic. "Well, most of us got off the moon," he commented to his wife in mindspace.

"Heinreich, I'm sorry about Charlie," Lise consoled her husband.

"I know," he said. "If I know Charlie, he was the one who detonated the hydrogen bomb. I should have known that he would have put others before himself instead of getting into a mindspace rocket."

"What is our status General Joos?" Heinreich asked a moment later as he settled back into his much larger conscious self.

General Joos projected the appropriate data to Heinreich's mind. "789 of us made it. Out of two-hundred Pratts, one-hundred-twenty made it back. One-thousand-thirty Wohlthats are four-thousand kilometers behind us, but, as we have learned, they can't catch up to us. If they light up their engines, we can do the same thing and best them."

"Can we then not open up a gap between us and them?" Professor Lacombe asked.

"We will, most assuredly we will," General Joos replied, "but only by so much as we have to conserve our fuel supplies."

"And if they launch missiles against us, I take it that we can easily destroy them with our long range excimer lasers?" Paul asked.

"Yes," Heinreich interrupted, "but conversely, if we try to fire missiles at them, they won't make it to their targets for the same damned reason. We are locked in a stalemate for now."

"A stalemate, lamentably, which will last until we get to Europa years from now," Lise added.

"That's a nice thought," Professor Lacombe said.

"Before we get too far away from Earth, shall we execute

Beta and rescue our orbiting refugees?" General Joos asked rhetorically.

"It's going to be tricky grappling those satellites without Charlie, but at least all of Primus One's fighters are following us." Heinreich stated.

***

"I have the first satellite locked," a Pratt fighter mindspace simulator called out to General Joos, who was leading the rescue mission. "They are ready to be grappled sir."

"Good Pratt 104," General Joos remarked. "Take it and get the hell out of here." The general was pleased with the rescue of the second satellite as well. So far things were going smoothly.

"Sir, this is Pratt 99, I have a bogey inbound for satellite number three which is currently flying over India. My sensors indicate it is a thermonuclear missile." Before General Joos could react to Pratt 99, he received a message from Pratt 21. "Sir this is Pratt 21, I have a bogey inbound for satellite number four which is currently flying over the Pacific. My sensors indicate it is a thermonuclear missile."

"Pratts 99 and 21, can you intercept the missiles and destroy them outside their kill radius?"

"Sir, Pratt 99 can comply."

"Sir, Pratt 21 is too far to comply."

"Damn it!" General Joos cursed. "Satellite four, how much fuel do you have left to change your orbit?"

"None sir," came back the satellite commander's reply.

"You know I can't get a fighter to intercept that missile. Pratt 21 is out of range. Do you have enough stored energy to discharge your laser?" General Joos pursued.

"Roger sir, we have enough for one shot."

The situation for satellite four looked hopeless. The satellite commander had one shot to destroy a thermonuclear missile careening towards him outside the kill radius of the missile. The situation for satellite three was much less dire. Pratt 99 simply had to destroy the inbound missile with its lasers outside the missile's kill radius, then grapple satellite three and leave.

General Joos could do little more than listen to the commander of satellite four counting down the seconds before he could fire his only remaining laser shot. When the count reached two, the general looked away and held his mindspace breath.

"We got the bastard sir!" the satellite commander broke in.

General Joos exhaled a deep sigh of relief. "Pratt 21 where are you?"

"Sir, Pratt 21 is inbound for a grappling pass."

"You do that damn it!" General Joos yelled out. "Pratt 99, what is your status?"

"Sir, Pratt 99 will engage bogey in two seconds."

After his second nuclear missile was destroyed by Pratt 99, Primus One did not launch any more missiles. They were too easy to destroy and too costly to waste. He promised himself, after the completion of his second fighter armada, that he would keep reserve fighter forces behind from then on. As large as his mindspace brain was, he had committed the error of an inexperienced commander. The price that he paid for his mistake was to watch a hundred Pratt fighters collect the two-hundred ALM spy satellites one by one to be attached to the tow lines of four large ALM rockets.

"Primus One, this is Major General Joos, see ya!"

***

A rag, tag fleet of one-hundred-twenty Pratt fighters and nearly ten-thousand eight-hundred mindspace souls pointed itself towards Jupiter. They were one month ahead of schedule, but this had not been by choice. They had left two-hundred of their kind dead on the moon, along with tons of precious supplies. Nevertheless, the survivors felt fortunate. They had rescued the ten-thousand mindspace souls living in the ALM spy satellites.

Once free of the Earth's gravity well, the mindspace beings switched off their engines to preserve energy—as did the remaining cloud of pursuing Wohlthat fighters—and coasted on Newton's first law towards the Jovian system along a smooth geodesic.

"When this trip is over three years from now," Heinreich commented, "the sun will look like a smallish, pale, cold disc from our new world. It will be a hell-of-a place to make a last stand for humanity. Any chance we can transition ourselves into the Europa manta rays before they are finished developing? Because if we can make it into the seas, the Wohlthats will be powerless to reach us from above."

"No," Lise replied bluntly.

"Too bad we can't order our lead cargo ships to fly faster to Europa and get there a month faster. That would fix things," Professor Lacombe commented. "But alas if we make them accelerate, they won't be able to stop themselves to enter orbit with their remaining rocket fuel."

"I know," Lise replied. "Charlie told us that in order to power all of our own personal mindspace server rockets, we would have no nuclear fuel to spare for the lead cargo ships. He had no choice but to design and build them with conventionally powered chemical rockets."

The remaining Wohlthat fighters were in no rush. They had caught the lunar base off guard, and had escaped with several captured ALM lunar mindspace survivors before the thermonuclear detonation. After interrogation and assimilation of the lunar ALM mindspace survivors, the Wohlthats—and Primus One—learned that they could bide their time. The fleet of ALM and SolDynetics mindspace lunar lifters would have to slow down once they got to Europa, and then spend a whole month loitering about while the mindspace manta ray beasts finished gestating in the seas of Europa. The remaining thousand Wohlthats, thus, would then have a whole month to annihilate the remaining Pratts, and then strike at the survivors of the ALM fleet at will. Once destroyed, Primus One could lay claim to the mindspace biota of Europa for himself, for later consumption at his leisure.

Ahead of the ALM refugees flew the conventional ALM heavy lift rockets that Charlie had designed. They were full of supplies that had been launched two months before the lunar attack. These autonomous vehicles kept in constant contact with their masters via standard EPR teleportation cryptography. Heinreich flew out to them a couple of times in one of the faster Pratt fighters just to make sure things were fine. He wanted to avoid surprises. Once he took a trip to recover spare parts for one of the damaged ALM lunar survivors, who's mindspace server was acting funny. Each of these roundtrips would take about a month. During these trips, Heinreich would think about his life and the history of the world he had witnessed. Heinreich marveled. His own father had ridden in horse drawn carriages in the 1890's and lived in cities without electricity. Now here was his son, only one family tree generation later, floating among the planets, participating in a fight for a new kind of life.

For the survivors of the lunar attack, aware all the time of the wolf pack following behind them, as the Wohlthats constantly taunted them over radio, the trip to Jupiter became a pensive and tense flight into the empty void of interplanetary space.

***

Every once in a while as the months passed by, Heinreich would lead a band of Pratt fighters back towards the remaining thousand Wohlthats. "If we had more nuclear energy stores, or a fusion reactor, we could plink them bastards off one by one and recharge our weapons," Heinreich told General Joos after one of his scouting trips.

The general, of course, knew that.

"According to my calculations, we can get about twenty percent of 'em maybe with what we've got," Paul added, "but in case you haven't noticed, the Wohlthats have split themselves into three staggered echelons so as not to have all their eggs in one basket."

Each time a small band of Pratts led by Heinreich approached the Wohlthats, one of the Wohlthat's would be sure to remind Heinreich who killed Donna and Karl, and each time without fail afterwards, Heinreich would pick out and destroy a Wohlthat before returning to the ALM fleet, spoiling the Wohlthats' plans to capture him.

Lise understood Heinreich's restlessness. She didn't like it though. "What would happen if you have a malfunction out there. The Wohlthat's would catch you or cut you to pieces." Lise, however, knew that her husband would never let that happen.

"You know I'd blow myself up before being captured."

Lise would then admonish him. "You know that I don't want to lose you again."

The fleet of refugees kept their main antenna arrays pointed back towards the Earth at all times. They wanted to keep tabs on events happening on the planet in general, and on any developments that might regard them in particular. Every few weeks the council of five—one less as Charlie was gone—convened to discuss the status of the fleet and its mission.

Paul Fritz brought a surprise to the group one day. With his secondary passive antenna array, he had spent his free time enjoying the wonders of the vastness of space by listening to the random radio chatter issuing from the local stars and beyond. At one point Paul thought he had identified a pattern issuing form the Sagittarius constellation that, *somewhat*, repeated itself. His first impression was that this *somewhat* repeating pattern might be the possible signature of an extraterrestrial intelligence, but he wasn't quite sure. Alone, he didn't have enough mindspace computer power resources to resolve the pattern's flow. Just when he would believe that he had it figured out, the pattern would change in a subtle way. When Paul let Lise in on his secret, she and he combined their resources and eventually did figure it out. The pattern was fractal, based on a relatively simple formula Paul had overlooked. That's why the pattern was self-similar but never repeating. When Paul and Lise announced their discovery to the rest of the group, it made for an exciting time.

"Mein God," Heinreich blurted out when he tuned his own passive antenna array to the spot Lise pointed out to him. "It is following the formula you figured out. To think there might be intelligent life out there one-thousand light years from here is shocking. What do you suppose it, the signal, might mean?"

Paul replied with mock cynicism that, "it might be a local sports broadcast from a wrestling match held a thousand years ago.

General Joos had a different idea. "They might be advanced beings who could be our allies one day. A thousand years ago these beings—unless we are listening to some natural phenomena—were spitting out radio waves. A thousand years ago the Christians were burning books and mankind was in the Dark Ages."

Professor Lacombe put General Joos' idea into perspective. "Joos is right to say we may be listening to some kind of natural phenomena. However, even if the signal is truly the signature of an advanced civilization, that signal was emitted a thousand years ago. Perhaps these beings destroyed themselves in the interim as we might yet do. If we survive though, and the signal persists, perhaps we might know the truth of things in a hundred thousand years, that being the shortest period we can manage with our current propulsion technology and gravity boosts to send a probe over there and back again with its report. Thus for now General Joos, the signal will remain an enigma that can play no role in our immediate future. We have to worry about getting ourselves into the seas of Europa somehow before the Wohlthats finish us off."

"What is the gap between the Wohlthats and us?" Lise inquired.

"About ten-thousand kilometers mein dearest," Heinreich replied. The refugees had slowly doubled the gap.

***

Another completely unexpected event caught the group of refugees off guard as they pressed on with their journey ever farther from the shrinking ball of blue they had once called home. They were six months away from reaching the edge of the asteroid fields when an enormous thermonuclear explosion lit up the sky about eleven-thousand kilometers to their rear, at just about the location where the rear Wohlthat echelon was flying thru.

"What the hell was that?" General Joos exclaimed as all eyes (sensors) turned back towards the explosion.

"I don't know," Heinreich fired back, "but just in case, I'm going to transfer myself into a Pratt and get the Pratt clone simulators out of hibernation."

General Joos thought that was a good idea. Paul, Lise and Pierre Lacombe began analyzing the absorption spectrum of the light and determined that many Wohlthat fighters had been evaporated.

"Something," Paul told the general, "has destroyed the rear Wohlthat echelon. The other echelons have broken up. Our radars indicate that the surviving Wohlthats are scrambling around in a very disorganized way.

"Something out there has got their attention," Paul added.

"Could it be an attack launched by some competitor to Primus One?" Lise asked. She sounded very concerned. "We didn't see any other launch facilities, but perhaps one of the other mindspace giants built such a facility beneath an ocean."

"That is a sobering thought," General Joos replied. "If it is, then we will have two forces after us."

Heinreich, now downloaded into his Pratt fighter, was itching to go out and take a peek, but knew better than to go poking around a fluid situation until he had better information to act on. His main duty—as always—was to protect the rag, tag fleet of mindspace survivors.

About an hour after the nuclear explosion, a frantic radio signal reached the ALM fleet. "Get me the hell out of here!"

"That was Charlie's voice!" Lise exclaimed.

In reality, it was a radio buoy Charlie had released that had transmitted the call for help before being vaporized by a Wohlthat. Without knowing this, hearing Charlie's voice was all Heinreich needed to go out and take a peek. "I'm taking ten Pratts and I'm going to get Charlie," he declared. "But what if it's a trick Heinreich? A trap? Look to what ends Primus One used the death of your second wife and child."

This didn't make sense to Heinreich. "Why would the Wohlthat's blow themselves up in huge numbers to trap me? That's nuts. Somebody or something out there—Charlie I hope—blew up over three-hundred of them Wohlthat bastards. I'm going to find out who or what it was that did this."

Charlie and seven other lunar mindspace beings were, in fact, in the fight of their life aboard a kludge of heavy lift nuclear impulse rockets. So far all of the false radar decoy buoys that they had deployed into space had kept the Wohlthats off their backs, but the supply of decoy buoys was running low. All Charlie could do—as he watched the heat signatures of all the Wohlthat fighters swarming around his spaceship chasing down decoy buoys—was to hope that help was on the way. "If I see the heat signature of one of our Pratt fighters, I'll fire a communications laser beam to it," he told his frightened mindspace server companions. "That will make sure the Wohlthats won't see us as a laser is a line-of-sight device."

Sure enough, the cavalry showed up—ten lousy Pratts against six-hundred surviving Wohlthats. It was better than nothing Charlie thought as he pointed his communications laser beam at one of the Pratt fighters.

"Sir, the bogey in question is here," the Pratt clone simulator that received the signal called out over its own secure laser beam communicator.

Heinreich received the coordinates of the bogey. "Let's go

get a closer look," he ordered. Heinreich also ordered five of his Pratts to break away and put some distance between themselves the bogey's path. When Heinreich and his five remaining Pratt escorts got within five kilometers of the bogey—it was an ugly thing made of several ALM rockets—they opened up a secure laser communications channel. "Can you tell me what the hell is going on Charlie?" were Heinreich's first words.

"What does it look like boss? We, seven of us, are trying to get the hell out of here!"

"Can you explain yourself better than that Charlie?" Heinreich asked. "I need to make sure that you're not part of a trap. I hope you understand my position."

Charlie did of course. "While the fight raged on the lunar surface, I came up with the idea of kludging together a few heavy lift rocket engines onto one rocket body, and then using the nuclear blast of plan Alpha as a means to sneak out. We blasted off twenty seconds before you called out for the detonation. Just when a group of Wohlthat's was going to finish us off, I heard you call Alpha, and we triggered the nuke. It toasted the electronics of the Wohlthats that were bearing down on us. We then hid in the debris field kicked up by the nuclear blast. For many weeks we let you and the Wohlthats fly on ahead of us. We figured the Wohlthats would be too busy keeping track of you guys, and not looking behind their backs. Slowly we tried to sneak up on them and give them a little gift, but the damned Wohlthats broke up into three huge echelons, lead, center and tail. We decided that before the Wohlthats broke up into any smaller groups we would take out their rear guard."

While Heinreich listened, he kept one eye on his passive sensors. The five Pratts he had ordered to break away had been detected. They were engaging a cluster of Wohlthats and doing their best to draw the enemy further away from Charlie's ship.

"So then what?" Heinreich asked.

"Well, my companions and I decided to take our chances. With our huge engines lit up, we whizzed past the rear echelon and dropped them a small package. You know, a little hydrogen bomb. After we blew the hell out of them, the center echelon of Wohlthats turned back towards our location as predicted. The EMP covered us for a little while. We took advantage of this to launch a whole pile of decoy buoys. That's when we called out to you hoping for the best. By the way Heinreich, we're packing five more H-bombs."

Heinreich paused to assess the situation. "While the decoy Pratts keep the Wohlthats distracted, I say you light up your engines and make a run for the fleet Charlie. With your massive engines they won't be able to catch you if you can get past their screening forces. Toss me one of you H-bombs. I'll catch it with mein grappling arm. Together with my remaining Pratts, I will draw off as many of the Wohlthats as I can. You have got to reach the fleet. General Joos can use the four remaining nukes to mine the asteroid field. It just might let us even the odds, or even beat the Wohlthats, because in case you haven't noticed, there are still over six-hundred left after what you did."

Charlie complied, lighting up his three engines to full throttle. "Good luck Heinreich," he called out as Heinreich and his Pratts broke away to distract the Wohlthats.

"We're over here you bastards!" Heinreich flagged the Wohlthats as Charlie's ship slipped away. "Come and get us," he continued taunting as he switched on his active radar sensors.

"Sir," Pratt 10 called out, "we have a group of two-hundred Wohlthats bearing down on our position. "Ahead of this group I detect twenty lead Wohlthats."

"Never mind the larger group Pratt 10. Let's chase down the lead echelon."

Pratt 11—which Charlie and Heinreich had tried to program with a more human personality when they had built their first batch of Pratt fighters—chuckled. "I was afraid you were going to say that boss, but what the hell, I've got nothing better to do out here. Do you?"

"As a matter of fact I do Pratt 11. I want you to start transmitting taunts using my voice. Keep on calling out, "I'm over here Peter," to draw the Wohlthats from me. I want to make a little special delivery to the leading echelon without being seen."

"Roger that boss," Pratt 11 replied.

The fight with the leading Wohlthats began with an exchange of laser fire. While Heinreich's five escort Pratts fought off the leading Wohlthats, Heinreich snuck away and pressed on to the front echelon with his nuke. At 10 kilometers from the lead echelon, Heinreich launched his H-bomb right into the middle of a pack of clustered Wohlthats. Those and all the other Wohlthats—who had been monitoring the fight against Pratt 11, thinking it was Heinreich—had not seen Heinreich's approach until it was too late. The Wohlthats did their best to break away, but over a hundred of them were destroyed when Heinreich triggered the nuclear detonation. Many of the surviving Wohlthats were badly damaged.

"Good," Heinreich thought. Only the middle echelon—along with a few lucky stragglers—remained intact, but that still meant nearly five-hundred-fifty healthy Wohlthats were still in the game. The ALM fleet, on the other hand, was now down to one-hundred-ten slightly beat up Pratts, making the ratio of bad guys to good guys five to one.

***

With Charlie's gift of four hydrogen bombs, General Joos thought that mining the asteroid belt was a good idea. He also thought that the remaining Wohlthat fighters would think the same thing, and informed the council of six of his expert opinion. "If we had a hundred nuclear weapons at our disposal like we should have had, I'd say let's use a few dozen of them to mine the asteroid field as we originally planned, but we only have four left. I also have little doubt that the Wohlthats are going to spread themselves out when they enter the asteroid field. At best, our bombs will only take out a small fraction of the bastards. For that reason I suggest we hang on to our precious few nukes."

"I have an alternative proposal," Heinreich stated. "I agree we shouldn't mine the asteroid field for the same reasons General Joos has just stated, but perhaps we can sacrifice two of our four nuclear weapons *just before* the Wohlthats penetrate the asteroid field. We can send a cloud of debris flying out at them shotgun style. I've asked Charlie to work on this scenario, and I will let him tell you what he came up with."

Charlie projected a video-like projection of his scenario into the mindspaces of his fellow council members. "Starting from the worst case to best case, here is what I determined. In the worst cast, using two nukes as a couple of shotgun shells against the Wohlthats before they enter the asteroid field will result in little to no losses of Wohlthats. But—and this is important—given that they will have to dodge all the debris flying at them, they will loose ground—maybe as much as two weeks. This will give the manta ray mindspace beasts two more weeks to gestate. As a best case scenario, we will not only cause the Wohlthats to loose ground, we might be able to take out a significant fraction of them. Either way, we gain precious time to let the sea beasts finish maturing. If we can get inside their bodies before the Wohlthats get to us, we'll be safe under the seas."

Paul quickly repeated Charlie's stochastic simulations and projections and liked what he saw.

"What Charlie hasn't mentioned yet," Heinreich broke in, "is that if we are willing to part with twenty Pratt fighters, we might be able to cause the Wohlthats even longer delays once they enter the asteroid field. The Pratts can make hit and run attacks with relative impunity in the asteroid field."

Lise didn't like this idea. "Of course, I take it that you would be leading the twenty Pratt clone simulators," she told her husband. "This scares me for obvious reasons, but we would also be reduced to having only ninety Pratts to protect the fleet."

It hurt Heinreich to think of Lise being left alone should he be killed, but delaying the Wohlthats was critical. "Lise, the two hydrogen bombs can buy the fleet two weeks, that cuts in half the time the fleet will be floating in orbit vulnerable to the Wohlthats, but delaying the Wohlthats by two weeks only makes our odds of success go from nearly impossible to just damned unlikely. We need to delay the Wohlthats even more. I think I can do that in the asteroid fields—maybe even gain us as much as two weeks. If I can do that, then our people will have mature manta ray mindspace bodies ready for them, safe in the seas of Europa."

Pierre Lacombe suggested that the council of six take the issue to the entire refugee fleet. "Our fate is their fate as well. Perhaps it is time that we start acting less like Primus One and more like a democracy."

After a few days of deliberation, the general consensus among the ALM refugees was to go forward with Heinreich's plan. Lise had suspected the outcome.

"Mein sweetheart, you know I will be fine," Heinreich tried to reassure Lise after the decision became official. Lise, for the first time in a long time, broke down crying in her virtual mindspace. She had a bad feeling in the pit of her stomach that she would never see Heinreich again.

"You know the first time I made love to you was about ninety years ago," Heinreich told Lise as he put his virtual arms around her virtual body. "Every time since then, mein love, it has always felt like the very first time."

Lise laughed a little. "Liar," she declared through her tears, but Heinreich didn't let her continue crying before he had her in his arms and was kissing her virtual mouth. For the next day and a half, while the nuclear bombs and Pratt fighters were prepared for a long duration mission, Lise and Heinreich made love in ways, and with intensities no corporeal beings would ever be able to comprehend.

"Whatever happens mein love, promise me one thing," Heinreich asked his wife after they collapsed into each other exhausted. "Promise me that you won't stay alone if something does go wrong. Your life might last millions of years or more."

Lise smiled, kissed the tip of Heinreich's mindspace nose, and nodded yes. She lied.

***

Heinreich watched with mixed emotions as the fleet of ALM refugees faded away into the asteroid field. "Okay boys. We have a job to do," he told his twenty Pratts. "In a little over a solar day's time the Wohlthats will near the periphery of the asteroid belt. That's when we think they will begin to spread themselves out, and that's when we'll detonate the two nukes and take our positions to bushwhack as many of them as we can."

"You bet your ass sir," Pratt 3 replied. Heinreich had liked the more human personality of Pratt 11. While the refugees had deliberated the mission, he and Charlie had augmented the twenty escort Pratts with the Pratt 11 personality algorithms.

It had been written a thousand times before, and Heinreich had experienced it himself a few times, the hardest part of war is the waiting—especially when the brains of the beings doing the waiting ran billions of times faster than they had in their previous wetware form. While Heinreich and his twenty Pratts waited, they checked the status of the two hydrogen bombs time and time again. The also studied candidate asteroids to best disburse debris. The final candidates were two large rocks, about the size of New York City. These asteroids were not very solid. Rather, they were highly fractured structures, and could be blown apart into countless fragments, each one, hopefully, a deadly missile. After the hydrogen bombs were planted, Heinreich and the Pratts ran trillions of stochastic simulations to model the effects of the two nuclear detonations.

"All we have to do now is hurry up and wait," Pratt 101 joked when the bombs had been checked for the thousandth time. "Shut up and check it again," Pratt 3 had replied. "The situation is normal—all fucked up," Pratt 33 blurted out.

The moment did come, finally, when Heinreich and his Pratts detected the heat signatures of the incoming Wohlthats. "Get ready to detonate the hydrogen bombs," Heinreich told his small company of escort fighters via his secure communications laser. "The minute they begin to break their formation is when we give them a little surprise. By the way boys, it has been mein honor to serve with you throughout this campaign. Bone Crusher himself would have been damned proud of you."

The moment the mass of five-hundred-fifty Wohlthats began to spread themselves out, they were greeted by two incredibly bright flashes of light and electromagnetic radiation across the spectrum. As predicted, the nest of hornets was dispersed into a chaotic swarm. As best as Heinreich and his companions could determine from their protected perches, perhaps as many as two hundred Wohlthats were flattened by the shower of missiles. Dozens of others likely suffered damage to various extents.

"I think we got 'em down to about three-hundred-sixty boss," Pratt 3 called out.

It went without saying that Heinreich wished things had gone worse for his pursuers. At least, he thought to himself, the remaining Wohlthats would have to proceed slowly and with extreme caution inside the asteroid belts. For one, to reduce the risk of having a large portion of their numbers destroyed by another nuclear weapon, the remaining Wohlthats would have to spread themselves out over a large volume. For another, the Wohlthats would be expecting hit and run raids.

Indeed, the first thing the surviving Wohlthats did after the two nuclear blasts was to spread themselves out. "Spread out like that boys, the Wohlthats are going to be easy pickin's as they penetrate the asteroid field," Heinreich told his Pratts. He was pleased.

Lise and the rest of the fleet saw the two flashes of the thermonuclear explosions. From where they were, the flashes of light weren't as bright, but were still impressive—enough so that Lise could only imagine how nasty the situation was for her husband and his faithful Pratt clone simulators.

Two weeks after the detonations, long range radar sounding confirmed that the Wohlthat fleet had lost about one week's worth of ground. Lise, though she knew that Heinreich and his twenty Pratts were by now engaging the surviving Wohlthats in hit and run raids, got hopeful that she would see her husband again soon. The nuclear detonation part of his plan had worked out perfectly after all. "I don't want to lose him again Pierre," she confessed to Professor Lacombe after the radar analysis was completed.

"You know your husband. He's a tough, resourceful man. He will pull through after delaying our pursuers by the amount of time we need, and at the last minute, he'll have himself teleported to a manta ray mindspace beast in Europa's seas."

Heinreich wished he could feel so confident. Over the span of the two weeks, he and his Pratts had destroyed twenty-one blast damaged Wohlthats at the cost of one of their own. With every hit and run attack, however, the Wohlthats improved their defensive tactics. One thing the Wohlthats never quit doing was to beam pictures of the mangled bodies of Donna and little Karl. The Wohlthats wanted Heinreich to lose his mind.

"Okay Pratts 33 and 34, let's close in on that Wohlthat. Looks like its missing a wing."

"Roger that," echoed the two Pratts as they armed their lasers.

Just about when Heinreich and his wingmen were about to fire on the damaged Wohlthat, three blinding flashes knocked out their sensors. "Mein god! The Wohlthats have detonated nukes of their own," Heinreich yelled out.

"Why the hell didn't we think of this happening?" Pratt 33 asked.

Sure enough the Wohlthats had attached deuterium/tritium spiked fission bombs to the backside of three asteroids. They had also planted sacrificial Wohlthats—purposely damaged—to draw in the Pratts.

The detonations spread a shower of deadly rock missiles in

the direction of Jupiter and the Pratt fighters. The Wohlthats weren't trying to strike at the fleet of refugees as they were too far away. Rather, the Wohlthats were trying to damage or destroy Heinreich and his band of Pratts, and succeeded in doing much damage.

Before Heinreich's saturated high gain sensors could come back online, a high speed boulder smashed into his Pratt fighter. "I'm hit!" Heinreich called out to Pratts 33 and 34. "And my engines are out."

"Hang on boss," Pratt 34 they called back. "Pratt 33 is smashed, but I'm going to get behind you and push you the hell out of here."

When the ALM fleet detected the three unexpected detonations, General Joos suggested the fleet light their engines. Charlie and Lise, performing a spectral analysis, determined that the weapons had been enhanced fission bombs—not quiet as powerful as hydrogen bombs—but still nasty enough at close quarters.

"It's got to be hell in there," was all Lise could think as she peered into the dying afterglow of expanding plasmas. "Can we send a Pratt fighter to go see what's going on out there?" she asked General Joos.

"What's one less fighter?" Charlie added.

General Joos didn't have a problem with the idea. "We definitely need to know what's behind us."

"Then I will go myself," Charlie volunteered. "Heinreich did that for me. I owe the boss that much."

Lise—sick with worry for Heinreich—didn't agree with this at all. "Heinreich went after you not only because you are his friend Charlie, but because as chief scientist with decades of experience in the defense business, you are valuable, far too valuable to lose."

General Joos agreed. "We will send out Pratt 7 to scout things out. Lets hope it gets lucky."

***

"I'm sorry boss. There are simply too many of them," Heinreich's wingman Pratt 34 called out just before it was vaporized in a barrage of laser fire. Heinreich was now alone among his enemies.

"Well Heinreich," one of the Wohlthats called out a second later. "I have to admit that you've fought a good fight, but it's over, and I will finish the job I started. But first—with my apologies—I'm going to have to board your fighter. You know how it is. Interrogations have their value. Who knows what useful secrets you may be harboring in that little mindspace mind of yours."

There was nothing Heinreich could do as a Wohlthat approached his damaged Pratt fighter. With his engines out, he could not even detonate himself. It was a grim sight seeing the Wohlthat docking itself to his Pratt body.

The Wohlthat chose an area towards the front of Heinreich's fighter that had sustained little damage. "You were lucky to have survived," the Wohlthat announced. "What had been your antenna array fixture, and is now only a pile of molten stumps, shielded your craft's nose, where your spintronic core housing is located. Any more damage and your mindspace would have been destroyed."

Heinreich didn't say a word.

Immediately in front of the stumps was a data port that appeared to be in working order. The Wohlthat, using its small robotic arms, gingerly docked itself to Heinreich's ship just past the data port and proceeded to connect itself to Heinreich's mind.

"Is this what rape feels like?" Heinreich wondered as he lay there dead in the water.

"Ah," the Wohlthat muttered as he pierced Heinreich's core neural circuits. "It's good to know your people only carry two more hydrogen bombs. Now stop trying to resist the data flow."

Heinreich had only one advantage in the situation. The Wohlthat clone simulator, despite that fact that its quantum computer core was powerful, did not have the power of Heinreich's more sophisticated spintronic brain. To spoil the efforts of the Wohlthat to pick his brains, Heinreich filled his mind with all manner of mathematical problems, and tried to figure if there was a deeper pattern to the alien signal Paul had detected months before.

"Are you trying to be a riddle wrapped in a mystery inside an enigma of spintronic thoughts?" the Wohlthat taunted Heinreich. "That is very brave of you Oberst von Onsager. But all I have to do is couple my CPU core to a few dozen of my clone copies to overwhelm your mindspace server's firewall defenses. So why not just cooperate, sit back and relax."

After seven Wohlthats coupled their CPU cores to overcome Heinreich's mindspace defenses, Heinreich tried distraction. "Why did you do it Peter? Why didn't you just kill me and leave mein family alone?"

The Wohlthats—programmed by Primus One—couldn't resist this question. "Because you didn't listen to me Oberst von Onsager. If you had followed my advice I wouldn't have been shot down and captured by the Soviets. And because while you thrived, I lived like a dog. You made no effort to seek me out after all the loyalty I had shown you, did you?"

Heinreich apologized. "You are right Peter. I did fail you. I had thought you were killed, but I should have tried to confirm this. So take your anger out on me if you must, but let what little is left of our humanity survive. Let my fleet live."

The seven Wohlthats burst out in laughter. "I couldn't do that Oberst von Onsager. If anyone taught me the meaning of duty it was you. By the way oberst, your Jew rat wife—that is what Hans called her isn't it?—is probably worried sick about you. That pathetic little fleet of yours dispatched a fighter to come looking for you. We destroyed it."

Poor Lise, Heinreich thought.

"Poor Lise indeed," the Wohlthats—listening to Heinreich's thoughts—repeated.

Heinreich continued putting up the best defense he could. He refocused his attention to the signal Paul had detected.

"Oh yes oberst," one of the seven Wohlthats that was infiltrating Heinreich's brain broke in. "We also detected the alien signal that is stored in your mind—if that is what it is—and informed Primus One. With his resources, he should be able to figure things out better than us. Now please stop resisting us."

Heinreich refused this request of course.

"Very well Oberst von Onsager, your resistance to our hacking is too strong. You thus leave us little choice. We will have to use nanobots, and we cannot guarantee that your mind will survive their onslaught. If you just let down your firewall and let us extract your secrets, I won't torture you. I will kill you straight out."

Heinreich refused this last request with obstinate silence.

One of the seven Wohlthats released billions of nanobot probes into Heinreich's spintronic mindspace core. Like little ants, the nanobots began to crawl all over Heinreich's mindspace innards. Every time a nanobot touched one of Heinreich's spintronic circuits, it would send a spark of bright light into

Heinreich's spintronic visual cortex, which manifested itself as pain. Before long, Heinreich felt as if he were floating at sea, trapped in a hurricane's fury, being crushed by hundred foot waves, whipped by winds, and shocked by tremendous bolts of lighting. No corporeal being would be capable of experiencing such misery, let alone survive it, and the pain only worsened. The deeper the nanobots probed into his spintronic core, the more memories, especially the bad ones, started to lash out at him. Heinreich started wishing for death more than he had ever wished for it before. "Please make them stop!" he begged his captors, but the process of extracting information from his spintronic core continued unabated. To add insult to injury, the Wohlthats transmitted the images of Heinreich's deepest, most personal thoughts to the fleet of refugees.

***

"We must go back and get him," Lise begged the remaining council of six. She was hysterical with grief. "Heinreich would do it for you," told them.

Charlie was tempted, but the rest of the council prohibited it.

"Heinreich knew that he was likely on a one way mission," General Joos stated. "That is why he had Paul purge all synchronization EPR teleportation secrets from his spintronic core before he departed," the general added. "Going back to get him would be suicide, and would squander away resources we can ill afford to lose. Heinreich wouldn't want that."

Lise, silent, couldn't argue against the general.

"We have to concentrate our efforts," General Joos continued, "on getting to Europa intact. The good news is that all of our supply ships arrived safely, and the ALM-Honda robots have nearly finished deploying all the staging areas.

The zygote seeding process is well under way. Thanks to Heinreich, the beasts, instead of being one month shy of being fully cooked, will be just a week shy of being ready to host our mindspaces."

Paul interrupted General Joos. "If everything goes well, and our remaining fighters can hold off the Wohlthats for a week, we should be able to entangle ourselves to the undersea beasts from low orbit around Europa."

"If we make it," Charlie added, "we should be safe as the Wohlthats are not built to operate in a liquid ocean. They can drop nuclear weapons on us, but they will be shooting in the dark. Europa is large and they won't be able to detect our undersea manta ray hosts."

Paul closed things out. "If we succeed occupying our host mindspace bodies, we must destroy any orbiting EPR teleporters. We don't want the Wohlthats trying to infiltrate our mindspace hosts."

Lise, though she understood the logic, didn't like it. "If we do this, Heinreich won't be able to transfer his mindspace to us from orbit."

***

"We are finished with you Oberst von Onsager," one of the Wohlthat clone simulators informed Heinreich. Heinreich was barely conscious when the nanobots were pulled out his spintronic core.

"My brothers and I have been debating what to do with you. About a third favor frying your core. About a third favor sending you crashing into Jupiter. That would be a spectacular way to go, and thus I oppose it. I, and about a third of my copies, prefer death by boredom, as we suffered in the Lubyanka gulag. When put to a vote, death by boredom was chosen." Heinreich didn't understand.

"We will disable both your radios and navigation thrusters and send you floating into the interstellar void. We will allow only your forward sensors to remain operational. You will spend about ten-thousand years staring into the jaws of infinity before your nuclear fuel runs out. We expect the loneliness will drive you mad."

After the Wohlthats tossed Heinreich into a deep space trajectory without any thruster control, Heinreich became, to all intents and purposes, trapped in a very cold and quite sensory deprivation chamber, the very vacuum of deep space itself. All he could see was the clusters of stars that lay between him and the Andromeda galaxy. For weeks with this awesome vista before him, Heinreich tried to repair his Pratt fighter's thrusters and radio equipment, but they were damaged beyond repair. All the while Heinreich drifted further into space, he worried about the fate of the ALM refugees.

Within a few months of sinking into ever deepening oceans of monotony, space delirium set in. Was there no Lise, was there no fleet of refugees, was everything a dream, Heinreich kept asking himself. The questions echoed in his vast spintronic mind in trillion-fold variations as his lone, forward looking sensor array peered out to the distant galaxy before his eyes. It would take him millions of years to reach the edge of the Milky Way, and a million times longer than that to reach Andromeda. No one would ever see or hear of him again. He had ceased to exist. He was dead, buried alive in the abyss of the universe. Several times when contemplating these thoughts, Heinreich broke down into what we would call tears. "I can't even shut myself off," he bitterly lamented. "For the next ten thousand years this bottomless pit will be mein life."

The profound silence crushed Heinreich's mind. His delirium worsened. Once he imagined himself in a Nazi uniform again. He was a general officer living in his home near the old Castle Academy by the Alps. The second world war was over. Hitler had won. Sipping a cognac in his back yard, Heinreich was enjoying a moment with Lise and Donna, his wives somehow fused into one woman. Outside, their two boys were playing with the family dog when several SS cars drove up to his home in the outskirts of Berlin. Soldiers in black uniforms and submachine guns alighted from their parked vehicles and proceeded to gather Donna-Lise and the boys. One by one, Heinreich's family were dragged towards his patio chair pleading for their lives, but to no avail. Each of them was summarily shot in the head while Heinreich looked on sipping his cognac. When it was over, one of the SS officers morphed into Hans. Shrugging his shoulders Hans remarked, "They were Jew rats you know."

"I know," Heinreich replied, "but such a pity in any case. Especially for mein son Karl. He turned four yesterday as a matter of fact. You could have shot him first to spare him the misery of seeing his mother and brother killed. Regardless, Captain Hans, you sullied my trousers with their blood—you see—the stain is right here." This imagined atrocity struck Heinreich hard. He dropped to his virtual knees beside his bloodied, murdered family and was filled with such powerful horror he wished to die for a moment, before realizing where he really was—billions of miles and a century away from Earth and its once and future Third Reich. The guilt he had carried with him all his adult life had a long reach.

On other occasions, Heinreich relived the car accident that took everyone but Douglas from him. With his superhuman intelligence, Heinreich worked out the physics of billions of deaths for his son Karl and his American born wife Donna. He would calculate the forces exerted on his Donna's head and determine when her aorta would rip away from her heart after her chest's impact against the dash board. Then Heinreich would calculate exactly when his son's spinal cord would snap, sending the decapitated body flying out of the back windshield of the car. These simulations drove him deeper into madness, but Heinreich couldn't help himself from running them over and over again as what seemed like centuries ticked by. He couldn't break away from his despair and his terrible, isolated loneliness. He wanted his children and his wives back, safe, in his arms, forever protected, or desperately to be dead like they were.

***

Seven years into Heinreich's voyage to infinitude, according to his internal clock, he was suddenly struck by a blinding light. His exquisitely sensitive forward looking sensor array went off-scale. Then there came a starless darkness even blacker and more profound than the one he had been buried in for 3.4×10^{43} conscious thoughts. It took more years it seemed before the tiniest pinpricks of light began to pierce through the darkness that had enveloped him. They were stars he realized after a period of dazed confusion, and their configuration was quite familiar to him. They were the stars the ALM fleet of refugees had used to navigate with, except there were also extra stars in his field of vision, stars that had been previously occluded by Jupiter. Heinreich realized that he was on the other side of Jupiter. This unexpected discovery absolutely dumbfounded him.

"My husband, it is I, Lise," a voice called out to Heinreich from the void.

"Is this a dream?" Heinreich asked. "I don't see how this is possible," he continued. "All of my communications devices are destroyed, thus I cannot possibly be having this discussion. I'm mad." Heinreich began to cry and laugh simultaneously. "I just want to die don't you understand? Please! End this."

Lise's voice continued to beam through to Heinreich's mind. "What you sense is real. It truly is I."

Heinreich remained quiet for a minute as Lise kept her thought presence a little distant. She didn't want to be pushy in her husband's fragile state.

"May I see you Lise?" Heinreich asked. A few seconds later the image of Lise began to form in Heinreich's mind. "I don't understand this," he stated after seeing Lise's image, still doubting himself.

"The last time you and I made love," Lise began to explain, "I secretly entangled myself with your being so that no matter how far apart destiny took us, in a way, we would always be coupled."

"What took you so long to find me? Seven years alone like no other being has experienced was killing me. I had only my guilt to live with, guilt for having been a Nazi, and for having caused Donna's and Karl's deaths as well."

Lise sighed. "My dearest, dearest husband, you have nothing at all to feel guilty about, and it was not seven years you drifted. The Wohlthats changed the settings of your internal clock. It was seven-thousand years that you drifted."

Heinreich was dumbfounded, but the surprises weren't over.

"It took that long for Paul and your son to figure out the meaning of the alien signal, and how this relates to finding you I will soon explain my husband."

"Mein son?" Heinreich asked confused.

"Yes my husband. The last time we made love, I mixed the DNA that you were, and that I was when we had bodies, in much the same way we built the Pratt clone simulators. "I named our boy Max, after my father."

Heinreich felt a surge of joy run throughout his spintronic core when he heard Lise's words. "When will I see mein son? Or is all this a dream?"

Lise calmed her husband down. "You will see him soon enough, but you are not yet ready. We are seven-thousand years evolved beyond you. Your spintronic core is far too feeble to comprehend the details of our new, truly non-corporeal reality."

While Heinreich digested Lise's incredible words, she projected a park bench by a placid lake into Heinreich's confused brain and asked him to take a seat beside her. "I can however tell you," she continued, "the story of the refugees for now, until we can bring your consciousness up to speed."

Heinreich was all ears.

"After we reached the staging platform orbiting Europa, we busied ourselves setting up the best defenses we could while our Europa sea beasts finished gestating. We knew the odds were very much against us. It was going to be a tight race against time. The beasts were one week too undeveloped, and the leading Wohlthats were only two days behind us."

Heinreich was pleased with himself. His planned had added three weeks time for the manta ray mindspace beasts to cook.

"None of us would have ever noticed the roiling seas of Jupiter if hadn't been for a near collision between one of our supply ships and an old NASA probe. Desperate for precious metals, we began to disassemble the old probe. That's when Paul and I, with our spintronic brains, noticed the coherent richness of Jupiter's atmosphere stored in the probe's database. No one had ever seen the coherence before because no one had ever applied the raw power of spintronic computers to the patterns before."

Heinreich was still confused as he stared upon the limpid waters of the virtual lake beside his wife. "I don't have any idea as to what you are talking about," he confessed.

Lise stroked her husband's virtual hair. "Jupiter's atmosphere had a self-organized coherence of organic molecules my dearest, a coherence organized in the way amino acids did on Earth when tossed into a jar of primordial ingredients, shaken, and given electric jolts. And the deeper we probed the gas giant, the stronger the coherence became. It was the answer to our prayers. We no longer had to worry about building ourselves a life as sea creatures on Europa or the threat of the Wohlthats. Jupiter itself would be our new home. Its matrix of coherence, we realized, would provide us a seemingly infinite spintronic mindspace infrastructure. We had won, and not just the battle for our lives, but for the Solar System itself."

Heinreich was even more confused. He thought he understood the Jupiter coherence, but not the winning of the Solar System. "How did you get rid of the Wohlthat fighter fleet to begin with?" he asked after he had picked up a stone and thrown it into the virtual lake.

Lise, still being cautious not to push her husband too hard, took his hands into her own and watched the spreading ripples for a moment. "We ignored them. Using Paul's small synchronization EPR teleporter, we moved our minds to Jupiter two days before the arrival of the Wohlthats. Jupiter—with its nearly infinite supply of raw mindspace material relative to the Earth—became our home. Each one of us refugees turned into mindspace super beings that dwarfed the combined mindspace resources of every living being on Earth."

Heinreich interrupted Lise. "What about Earth? What happened?" He expected to hear bad news.

"As for the Earth, ignoring the puny armada of Wohlthats that began searching for us in vain in the seas of Europa, we immediately went on the offensive against Primus One. We started by entangling ourselves with our heavy lift nuclear impulse rockets and our two nuclear weapons. We sent the nuclear tipped rockets back to collect small asteroids, and had the rockets entangle themselves with their cargo. After that, we sent the loaded heavy lift rockets towards Earth. About five times outside the radius of the moon, we detonated the two weapons spreading out a cloud of entangled asteroid dust several times larger than the Earth on a collision course with the Earth. When the entangled asteroid dust struck the Earth's atmosphere and began floating down to its surface, we began using Paul's EPR synchronization teleportation technology—which now had access to the virtually unlimited amounts of energy and computing resources of Jupiter—to grow an exponentially coherent entanglement between ourselves on Jupiter and the Earth. Within a few weeks we began entangling our minds with everything in Earth's atmosphere and surface, and its seas. No mindspace entity on Earth suspected a thing, and before too long, we turned our far superior mindspace capacities to ending the war by breaking past the feeble firewalls of the carnivores and cleaning out their minds. I'm sorry to say that of the ten-billion that lived at the outset of the mindspace wars, only three-million survived, and most of these survivors were in zombie states. We had to peel away the Trojan horses from out of them."

"So now tell me about me. How did you find me?" Heinreich asked.

Lise lowered her eyes. "After I entangled myself with you, you and I would be forever linked. The problem was that I couldn't use Paul's EPR teleportation synchronization technology on you because you were simply too distant and too small an object to be found. We searched for you, but you were a needle in an interstellar haystack. The more time that passed, moreover, the weaker our link got as noise from the vacuum was always deteriorating it."

Lise paused to let Heinreich speak. "So how does mein son and Paul relate to you having found me?"

Lise projected an image of the star system from where the alien signal had issued into Heinreich's brain. "Do you remember the alien signal we detected on our way to Europa?"

Heinreich nodded yes.

"It is a long story—I'm not trying to be facetious," Lise began with a smile. "As the years passed by, both your son Max and Paul continued to study the alien signal we had detected. After about five-thousand years of working on it, the two of them concluded that it was definitely artificial, and contained clues on how to tunnel out of our universe and into other universes separated from us by energy barriers of various scales. It took us another thousand years to develop the technology, and we had to harness nearly the entire energy of the sun to do so. We then spent the next thousand years leaping about from universe to universe. The vast majority of them were barren, as their properties precluded the evolution of organic life. We mindspace beings could survive in them because we had learned how to convert ourselves into string-like resonances of space-time itself. A very tiny portion of the universes we found had properties in which life could evolve. As we explored more universes, we began to realize that all the universes we had visited, as well our own, were immensely small along a new dimensional variable. Once we realized this, finding you became child's play. And here I am my husband, happy to have found you before your nuclear fuel ran out."

"But what about the aliens who left us the signal?" Heinreich asked. His curiosity was piqued.

"We've never found a trace of them," Lise answered.

Heinreich's mind was boggled when heard this. He told Lise he needed to rest his circuits.

After a few microseconds of rest, Heinreich turned his eyes to Lise. "Do you know what I'm thinking right now Lise?"

Lise could see that Heinreich, presenting himself as a young man in virtual spintronic space, had a grin on his face, the very same grin she had learned to adore so many millennia before.

"Tell me my dearest," Lise replied with a warm, happy smile of her own.

"I'm recalling," Heinreich began, "the paradox of Colonel John P. Stapp that he stated in reply to Murphy's law back in the days when I was flying at Edwards Air Force Base. Captain Edward A. Murphy was working on Air Force Project MX981 when he stated that if anything can go wrong it will. Bone Crusher and I got a good laugh out of it when we heard it, but when Stapp, who was our flight surgeon, and the very same idiot who rode the deceleration rocket sleds in the 1960s, heard it, he stated without skipping a beat that, *the universal aptitude for ineptitude makes any human accomplishment an incredible miracle.*"

Lise laughed a good laugh. For all the human and trans-human knowledge that she had gathered and catalogued, she had never come upon Stapp's Ironical Paradox. Lise's virtual representation then placed its hand on Heinreich's virtual cheek. She had tears welling in her eyes. "I love you," she told him as he moved her virtual form into his arms. Heinreich once again looked upon the face of Lise, radiant, full of life, glowing with happiness. She was as beautiful as the first time he had laid eyes upon her as a child, as well as when she had fled from Germany, when she had smiled for him as bravely as she could before leaving for Paris. It was the year 9053, and Heinreich was filled with joy. So too was Lise at long last. "I want to meet mein son, Max," Heinreich told his radiant wife.

EPILOGUE

A weakness of humanity that was retained by its distant post-human descendents was its innate competitiveness. It had to be so. Life evolves. In the animal kingdom it does so by passive accident and selection. An inextricable part of intelligent evolution, however, is the active seeking for unexploited resources and niches. Whether the seeking is done out of pure curiosity or for the purpose of active exploitation does not matter. When a novel resource or niche is discovered, it is done so by some partial portion of a civilization. Given the finite rate at which information flows, it takes time for the information concerning the new niche to diffuse to the rest of the given civilization. Thus before the information can be diffused for the benefit of all, there is some finite, minimum period of time in which some partial portion of the civilization has a competitive advantage over the remaining portion. To some degree, there thus arises a wealth gradient between the haves and the have-nots, and thus there arises the potential for class conflict, especially when the flow of information is purposely impeded. Think of the value of maps placing the location of rivers and mountain passes and so forth when the New World was being explored by the old, competing European powers.

As the Communist Soviets learned in their attempts to force homogeneity on their people, without competition for resources, motivation dies. Somewhere between complete anarchy and extreme homogeneity, life thrives only when competition is allowed a great, but limited degree of freedom—laissez faire. Unfortunately maintaining a stable, balanced equilibrium is not easy. Envies and rivalries arise from competition, sometimes to the point of leading to heinous crimes on the individual level, and to crimes against humanity on the national (or trans-galactic) level. Maintaining the balance requires a continuous effort and great wisdom.

Given that wealth gradients must exist, humanity's descendents worked hard at not allowing extreme wealth gradients, such as the ones that grew on Earth—individual billionaires rivaling the wealth of billions of poor, colonial exploitation by empires and the such—to grow. To address this issue a body politic—a representative government with checks and balances—was built. After weak participation, attendance of the sessions was made mandatory for all. A benefit of this was that it allowed a census to be taken. From cosmic eon to cosmic eon, for inexplicable reasons, members of the post-human society vanished from the planes of existence accessible to them. To where and to what end the disappeared ones vanished, no one knew.

Other traits of humanity, though they stubbornly resisted, eventually died off. Two of the first such features to go after life spans became indefinitely long were the need for faith in the existence of a supreme being and the concomitant need to practice dogmatic, artificially ritualized religion. When death was still possible—when trans human beings were merely tucked away in redundant mindspace servers, some still felt a need for gods, lest a comet destroy the Earth. But when the offspring of humanity took on their forms as resonant structures of folded space-time and tunneled to other universes, religion and the gods perished.

It was not an issue of proving or disproving the existence of an omnipotent, omniscient god or gods. This is impossible. All finite beings, no matter how astronomically advanced their knowledge base may be, are still finite beings. The supply of new universes, for instance, seemed infinitely inexhaustible to the offspring of humanity, but no one could prove this. Rather, the issue with belief in a god or gods became an issue of choice using Occam's razor to decide between drunken, fear-based absurdity and lucid, cold rationality.

In its simplest form, Occam's razor states that explanations should never multiply causes without necessity. When two explanations are offered for a phenomenon, the simplest full explanation is preferable. It was incomparably simpler for humanity's descendants to believe that life was nothing special—just something that happened every once in a while in certain universes out of a greater majority of barren universes—than to believe that life was the product of an eternal, perfect, omniscient, omnipotent being. The latter belief system, far from being simple, was riddled with an inexhaustible supply of logical absurdities. Can Hitler be faulted when it was God, who long knowing the future horrors to be unleashed, went ahead an created the dictator? Or does the buck stop with God? A belief in such a god also came with an innately pernicious requirement, namely, the requirement for faith. Faith in a belief system is tantamount to having no doubt about the belief system, such as having no doubt when committing atrocities in the name of one's gods.

Looking at human history, only under a moral philosophy of "no doubt" could entire hordes of religiously motivated people throughout the human ages, by reason of their "no doubt" faith, have become Teutonic Nazis, or members of the KKK, or of Al Qaeda, to practice witch burning, lynching, misogynist sexism, honor killings, homophobic crimes, child molestation, and so many other types of nefarious crimes against humanity in the name and will of some god. "No doubt" gave the faithful the right to raze entire civilizations, pillage, plunder, murder, maim, destroy, burn books, imprison scholars, discriminate, rape, butcher, segregate, and slowly eviscerate other peoples.

Even when religion was doing good deeds, it was a costly, resource wasting middleman. All contributions from religion to charity and to community, to medicine and so forth, were but mere fractions of the collected, tax-exempt monies. Had the middleman of religion been cut out, much more of the peoples' charity could have gone to the search for cures. How many died because of lack of research funds for cancer alone? And how many of the those who did charity work in the name of their god did so out of true sincerity, versus those who, out of self-interest, did so to get into their respective ideas of heaven and promised lands?

The descendents of humanity rather adopted a philosophy based on an eternal doubt of everything according to the scientific method. All "truths" were recognized as being merely models valid in certain ranges of applicability, as Newton's law of gravity was decent enough for plodding along in the Solar System, and Einstein's theory of general relativity was decent enough for plodding around galactic clusters, and so forth. By giving up faith, and doubting everything, the decedents of humanity became less of a danger to themselves, as holy war without faith is difficult to promulgate. The descendents of humanity found meaning in their lives, and in what they did for each other. Their hypocrisy was open. Good deeds and the pursuit of communal harmony were done not for the rewards of heaven, but because such behavior was necessary for the better good and stability of the entire civilization, and thus for the better good of the individual. Many atheists on Earth lived fairly much like, say, good Christians. They did so not for eternal rewards or out of fear of being eternally damned by a so-called god of love, but out of fear and/or respect for the secular laws, and for the good of their communities, and hence for the good of themselves and their loved ones. With cooperation there is a net gain for all.

When looking back to avoid old mistakes, the children of humanity could not believe that many of their ancestors could have been so brainwashed by religion and faith in a god. In Judeo-Christian-Muslim type religions for instance, the faithful were asked to believe that an eternal, perfect, omniscient and omnipotent god, who knew an eternity before creating humanity, exactly what would happen after he created humanity, namely, that humanity would mess things up, would punish the wicked and reward the good. Given this god's omniscience, the so-called wicked were thus, by means of predetermination, condemned to their fates an eternity before they ever saw the light of day. The god itself, that supposedly initiated all existence—and knew the future—was never to blame, and never seen as a masochist with an ego so weak it demanded worship from its lowly creations. These faithful, at least the majority of them, never even had the curiosity to ponder the logical absurdity of whether an omnipotent being could create a burrito so hot-n-spicy that even he couldn't eat it.

Those faithful who did confront the logical absurdities of their faith typically relied upon the combined "father/pea brain" analogy in the defense of their faith. *When you were a kid*, they would tell the atheists, *and your father denied you ice cream as a punishment, he was doing it for your own good, to protect you and to teach you a lesson. As a child you could not have understood your father's logic, and you probably thought he was being bad to you. In the same sense,* they would argue, *our god is our father and we are its children,* and justify in this way the existence of good and bad. *Without bad one can't have good, and that's why we have bad in our god's world, so that we may learn and appreciate things.* But what good ever came of genocide? What lesson did the annihilated peoples, the children, mothers and fathers, learn? What benefit was conferred when a five year old was molested by a priest, died of cancer or was shot in the back by a Muslim terrorist? Did that god have to preordain the creation of a doomed child to teach the child's parents a lesson? Or to help pay for an oncologist's shiny sports car? Or because that omnipotent god needed the kid's help in heaven as some weeping mothers would say?

Such illogical absurdities, infinite in number, were just washed away in the pea-brain defense of god. *God works in mysterious ways* was the fall-back position of the faithful. We, tiny little pea-brained mortal children of God cannot understand Him. But if the children of the god were too pea-brained to understand their god, how could they possible worship their god properly? And why would He need their miserable, infinitesimal worship? Out of infinite ego? By saying that God works in mysterious ways, people of faith were avoiding the ultimate question of their faith: why does the buck not stop with You? This is not a hard, arrogant question. It is a simple question. The answer should not involve the higher mathematics of a god that would dwarf an Einstein. The Nobel Prize winning physicist Richard Feynman—who said that if you can't explain something to your grandmother, you don't understand it—would have likely agreed.

Thus, between being nothing special in a sea of infinitely

many possible realities and faith in an omnipotent, omniscient, eternal god of love who blames the miseries and faults of existence on all but himself—hence making himself the greatest enemy to human life and dignity, law and order— the offspring of humanity chose to believe they were nothing special in the grand scheme of things.

An eternal, perfect, omniscient, omnipotent, eternal being is a dead lump of nothing that would have zero motivation for doing anything. Create, or do anything, but what for? It would know the outcome an eternity beforehand, hence it would have zero motivation to act in the first place. Finite beings, no matter how advanced do not suffer this logical absurdity. When finite beings procreate, they can have no perfect vision of the future of their progeny. Lacking omniscience allows for the possibility, if not the guarantee of motivation. This was yet another fundamental distinction that humans of faith never seemed to realize.

In the end, the great nineteenth century thinker Immanuel Kant was wrong. Even before the outbreak of the mindspace wars, millions of atheists existed practicing the same morals of most religious peoples, as in not stealing, cheating, killing, doing unto others as they would have done to themselves, and so forth. Morality—which has more to do with self-interested actions for the net gain of self and civilization—needn't have anything to do with dogmatic, pernicious religious faith in logical absurdities.

Max von Onsager

ABOUT THE AUTHOR

Alex Alaniz, Ph.D., is Los Alamos-based nuclear weapons physicist, as well as a pilot. A former military officer, he has expertise in molecular dynamics, space-based weapons, financial engineering and operations research. In the epic styles of Tolstoy's War and Peace and Clarke's space odyssey, his historically based vision portends the dangers of the fast approaching post-Darwinian age through Lise and Heinreich. Lovers divided by Hitler's war until reunited in twenty-first century quantum mindspace servers, Lise and Heinreich must fight to preserve post-humanity's children. With a background in molecular dynamics, weapons and operations research as a military officer, and cut-throat energy market financial engineering, Alex's fascinating story warns us—our innate competitiveness may soon spell the end of us.